PRINCIPLES AND METHODS OF ANIMAL BREEDING

Principles and Methods of ANIMAL BREEDING

R. B. KELLEY, D.V.Sc.
Animal Geneticist,
Council for Scientific and Industrial Research;
Officer-in-Charge, the F.D. McMaster Field Station

AXIS BOOKS (INDIA)

Published by:
AXIS BOOKS (INDIA)
10-Vyas Ji Ka Nohra, Sardarpura
Chopasani Road, Jodhpur – 342 002
Phone: 91-0291-2642319, Fax: 2643993
E. mail: agrobios@sify.com

AXIS BOOKS (INDIA)

ISBN (10): 81-909065-2-6
ISBN (13): 978-81-909065-2-4
***Price:* Rs. 990.00 / US$ 66.00**

Published by: Dr. Updesh Purohit for Axis Books (India), Jodhpur
Lasertypeset at: Yashee Computers and Printers, Jodhpur
Cover Design by: Reena
Printed at: Bharat Printing Press, Jodhpur

INTRODUCTION

In conjunction with the Sydney University Veterinary Faculty, the New South Wales branch of the Australian Veterinary Association is conducting periodic refresher-courses for its graduate members and for officers of the Agricultural Department. The subject for one course will be Animal Breeding.

The President's invitation to conduct it presented me with an entirely new outlook on the subject. Lecturing is not a regular part of my work, and it was therefore necessary for me to construct from the beginning a course that could be covered in the two weeks of the proposed school.

While thus engaged I realized that, if it were possible to put forward a concise story of animal breeding as the basis of the lectures, the work would probably be of interest and service to many breeders of livestock who would be unable to attend.

With this end in view, the subject has been discussed as a statement of principles and methods. I hope that appropriate levels of simplicity and of discrimination have been adequately maintained to associate established practices with biological precepts for the information of breeders. Readers who so desire will find discussions amplified considerably by works quoted in Appendix III and also in textbooks by such authors as Fisher, Rice, Nichols, Lush, Winters, Hammond, Babcock and Clausen, or Waddington, to mention only some that are available.

A number of pedigrees have been included in early sections, but opportunities were not available to check them with original records. They should therefore be regarded only as descriptive of methods stated in various breed histories. Symbols, such as those used frequently in later sections, would have provided alternatives and, indeed, they did in some instances; but, because I believe the animals' names to be substantially correct, I used many. However, I did not include registered numbers which would have made the statements irrevocable.

I am very appreciative of assistance the staff of this Field Station has given in preparing the work; especially Miss E. M. I. Brown, B.A. and Mr H. G. Turner, B.Agr.Sc., who read and checked the manuscript, and Miss S. E. Boult who prepared the typescript. To them, to the many authors whose work provided the background of the discussion, to *Country Life* for which the greater parts of Chapters XIV and XV were first written under a nom-de-plume, my sincere thanks are extended.

R. B. K.

The F. D. McMaster Field Station,
Badgery's Creek,

CONTENTS

CHAPTER		PAGE
I.	HISTORICAL BACKGROUND	1
II.	ROBERT BAKEWELL	9
III.	SHORTHORNS	15
IV.	SHORTHORNS—(*Continued*)	27
V.	ABERDEEN-ANGUS	38
VI.	HEREFORDS: HORSES: REVIEW OF CONSTRUCTION PERIOD	49
VII.	BIOLOGY	59
VIII.	MENDELISM	72
IX.	POLYGENIC OR MULTIPLE-FACTOR CHARACTERS	91
X.	PHENOTYPIC SELECTION	105
XI.	GENOTYPIC SELECTION: ISOLATION	120
XII.	MODERN ANIMAL BREEDING	144
XIII.	ANIMAL BREEDING IN AUSTRALIA	177
XIV.	ARTIFICIAL INSEMINATION	210
XV.	PROGENY-TESTING OF FINE-WOOL SHEEP	228
XVI.	CONCLUSION: CURRENT RESEARCH	244
	APPENDIX I: WRIGHT S COEFFICIENTS OF INBREEDING AND RELATIONSHIP	261
	APPENDIX II: GLOSSARY OF TERMS	273
	APPENDIX III: REFERENCES TO LITERATURE	283
	INDEX	289

CHAPTER I

HISTORICAL BACKGROUND

DOMESTIC animals are repeated samples of the composite heredity of their ancestors. Their ultimate origins are obscure, but they certainly antedated written history; and domestication also took place at a very early date. Since then, environmental conditions for their survival have been modified by artifacts and by man's peculiar needs. In England, for example, wariness and fleetness in cattle, upon which their safety depended under natural conditions, were reduced by protecting them from predatory animals many years before food was stored for maintenance during the winter. Similarly, although horses native to England were suitable for many light duties, iron-clad knights of medieval times imposed weight-carrying as a further condition for their usefulness and survival. To satisfy the new requirement ponies were mated with Continental horses of greater size, and some lost their identity.

Thus from earliest times, the environment of domestic animals had two aspects. One was inherent in their geographical situation; the other resulted from their association with man. These, together with the heredity of the animals, were the forces responsible for existing breeds and strains. Specification of livestock as indigenous to any country, therefore, must have reference to particular times.

Celtic Shorthorns and Early Britons were contemporary. The cattle are now represented by more or less improved types in the Highlands of Scotland and in

Kerry. Probably they should be regarded as truly indigenous to Britain.

Improvements effected by the Romans were dictated principally by military requirements and took place mainly in the neighbourhood of their camps. They introduced from the Mediterranean littoral horses and cattle used in the trade of war. Wilson (1909) states that these oxen, mated with Celtic Shorthorns, were progenitors of the White Park Cattle of England. In support of this he presents evidence to show that all such herds were in the neighbourhood of pre-existing Roman towns or camps.

After the Roman evacuation, Angles, Saxons, and Jutes alike despoiled and displaced Britons, driving them into the more inaccessible parts of Wales, Scotland, and Ireland. Invasions by Norsemen were not hit-and-run raids; whole communities migrated and settled in the conquered territories. To this day there are two schools of thought with regard to the origin of polledness among cattle in the eastern counties of England. One traces the condition to indigenous strains brought from the south of Scotland, while the other considers polled cattle to be descended from stock introduced during invasions from Norway, where native polled cattle still exist.

When conquered by the Normans, England had its population distributed in a series of crude agricultural communities isolated from one another by forests and moors but holding together for mutual protection. Each was but a slight advance upon the tribal life from which it sprang.

At that time Celtic Shorthorns, with two modifications, were the cattle in vogue. Few of the primitive forest and moorland ponies were in general use. There are no records to show where the sheep came from; but apparently there was sufficient locally grown wool to supply material for the woollen manufactory established in Winchester by the Romans, for it is unlikely that they would have brought the raw product to Britain. However, sheep are not specifically mentioned until A.D. 712.

Conquest by the Normans and their subsequent administration of the kingdom were characteristically feudal. The unit was the manor-house surrounded by its village. Each was self-contained. There were cultivation strips and grass paddocks, both of which were common property. Oxen were used principally for draught purposes, and the level of nutriment during winter was so low that surplus stock were slaughtered and salted down in the autumn.

After introductions of cattle by the Norsemen and horses by the Normans, no outstanding additions appear to have been made to the genetic constitution of domestic animals in the British Isles until cattle were brought from the Netherlands and Arab horses from the East. However, great secondary environmental changes were taking place. There was, of course, no primary or climatic change, but man's artifacts were improving and his needs were expanding. Broadly, these constitute the history of animal husbandry in England prior to the eighteenth century.

There were three outstanding changes in the secondary environment. The first was the enclosure of common land. This movement resulted in farms of greater size on which control by individuals replaced that of communities with communal rights. Until the old system of common grazing was abandoned, isolation of groups of livestock and their improvement by use of selected sires was impracticable. The second change took place with improvements in agriculture, notably introduction of the Teutonic plough which turned the sod. It replaced the scratch-plough of Roman origin which threw the soil on both sides of the furrow. Improved methods of cultivation were accompanied by the sowing of turnips and clovers. The third change was the rapid growth of industry which concentrated and employed relatively large human populations in towns and cities. This created an unprecedented demand for animal products, particularly meat.

ENGLAND AND SCOTLAND

Surveys made of animal populations up to the end of the eighteenth century in Great Britain indicate that there were many distinct groups among all kinds of livestock. Youatt (1837-74) described different groups of both sheep and cattle in practically every county. Evidently the truly indigenous stocks remained almost pure in remote parts of Ireland, Cornwall, Wales, and Scotland, but every other group had been influenced by greater or lesser admixture with intruders. Such crossbreeding may or may not have been deliberate but, together with the isolation of groups resulting from indifferent transportation and local patriotism, it undoubtedly had far-reaching effects.

The crossbreeding provided a fundamental store of heterogeneity which facilitated selection, and the isolation of groups, whether geographical or deliberate, as is possible for selected animals, enabled distinct types to become established. Some of these had reached quite high standards of productivity during this anonymous period of breeding. Thus Holderness and Teeswater cattle in Durham and Yorkshire, which had been influenced very largely by Dutch infiltrations, were notorious for high milk-yields and considerable size. Youatt states that they frequently yielded twenty-four quarts of milk daily, and according to Sinclair (1908) a seven-year-old beast, killed in 1779, yielded two forequarters of beef weighing 75 stone 7 pounds, two hindquarters weighing 76 stone 3 pounds, and tallow 11 stone; a total of 162 stone or 2278 pounds.

The cattle of Herefordshire also had a considerable admixture of Dutch "blood". They had been developed principally for draught purposes and as grazing cattle; and as such they enjoyed a high reputation. Although Dutch cattle, and also groups of English cattle, were taken to Scotland, apparently those that gave rise to the Aberdeen-Angus breed were more nearly indigenous than the progenitors of Shorthorns or Herefords. On the other hand Ayrshires, which are also Scottish cattle, have a very heterogeneous ancestry, in which both Dutch and Holderness-Teeswater cattle appear superimposed on the native Kyloes.

Although sheep in different localities were dissimilar, there is no evidence that attributes their differences to an admixture of foreign races. It is possible that the Romans may have introduced their fine-wool sheep and so modified some of the short-wools. However, it is apparent that for a very long time the most distinct difference was between the short- and long-wool breeds rather than between strains of either. Quite possibly differences within these could be attributed to their isolation. Sheep did exist at a very early date as two kinds, each of which had distinct sub-groups, but little more has been established.

Horses had developed from various foundation stocks, probably cross-mated, so that breeds varied in type from small Shetland ponies, through a wide range of larger ponies and cobs to light or massive draught-horses. The English Thoroughbred owes its speed and conformation principally to infusions of Arab blood. It is suggested also that some of the ponies and probably most of the coaching sorts had a similar ancestry.

Breeds and strains of livestock that were present in Great Britain in the eighteenth century are sometimes regarded as indigenous. All are not so in the literal sense. It can be shown why they arose, but their precise origin or how they arose or who was instrumental in the process cannot be determined. However, a number of kinds were deliberately developed from them and these, collectively, are known as British breeds of livestock.

NETHERLANDS AND CHANNEL ISLANDS

Prentice (1942) has made a close study of the early history of breeds of dairy cattle. He considers that those characterized by low fat-content but high milk-yield and those with opposite characters are the two foundation kinds. He strongly suggests that the first, which appeared earliest in

North Italy, moved from there to become progenitors of the large Dutch cattle and so of Shorthorns and Ayrshires. Those with relatively low milk-yield and high fat-content he considers were brought south by invading Norsemen. Their progenitors were a distinct race of polled cattle that moved north from Asia in prehistoric times.

He shows that Channel Island breeds have a heterogeneous ancestry that was compounded relatively recently. Beginning with a small black beast, counterpart of the Celtic Shorthorn, modification first took place in the ninth and tenth centuries as a result of crosses with cattle brought southward by invading Norsemen. In the middle of the eighteenth century the descendants of these were again crossed, this time with Dutch cattle. About the middle of the nineteenth century, after the Napoleonic Wars, Shorthorns and Ayrshires were introduced in addition, and possibly Brown Swiss cattle also. The breed now known as Jerseys traces, he states, from these cattle at a period about 1850. The Guernsey breed is apparently more closely affiliated with cattle on the mainland.

MERINOS IN AUSTRALIA

Youatt states that, until the introduction of turnips and clovers to English agriculture, British short-wool sheep had supplied fleeces suited to the manufacture of broadcloth. After that time wool from Spain was better suited for the textiles then coming into vogue. Spain had had fine-wool sheep for centuries. He attributes their origin to sheep brought to the peninsula by the Romans. Another suggestion commonly made is that Spanish sheep were crossed with rams from North Africa. Be that as it may, there were large numbers of fine-wool Merino sheep exclusively in Spain before the opening of the nineteenth century. Their exportation was prohibited, but eventually they reached Prussia, Saxony, France, and America. A few went to England and some, directly or indirectly, to the Cape of Good Hope.

When Australia was first settled by Europeans there were no indigenous animals suitable for food, and every supply ship brought what it could from its ports of call. By these means Cape and Indian sheep and British long- and short-wool kinds arrived in the colony, where they reproduced freely.

In course of time the captain of a supply ship at Capetown purchased, as a private venture, about thirty head of Merinos. Many were lost on the voyage but he was able to land some, probably about ten or twelve, in Australia. These were bred in isolation and by crossbreeding with the stock already in the colony. The value of wool from them and from their crossbred progeny was realized in England and in Australia. This was the beginning of our fine-wool industry.

Immediate expansion of numbers took place, and Merino sheep were imported from all the oversea centres to which they had spread. They were few, however, in comparison with the rapid increase of the fine-wool flocks. Further, it was generally found that the imported sheep were unsuited to the whole of the country. Under these conditions two flocks became outstandingly important. Their owners, using the foundation sheep that were available, in the words of the elder Peppin, who was one of them, "found the type of sheep that suited the country and developed it". There can be no reasonable doubt that in doing so both these flock-masters, Murray and Peppin, began with heterogeneous sheep.

In this respect their flocks were characteristic of foundation animals of all domestic livestock. In every case many breeds were thrown, often by chance, as ingredients into a number of melting pots. From these, as circumstances arose, samples of the composite mass were drawn as specific breeds and strains to satisfy contemporary requirements.

Heterogeneity in foundation animals provides a sufficiently wide range for selection, but economic or secondary environmental causes are the principal controlling factors in the evolution of breeds and strains of domestic animals.

These are still critical. They can be traced quite clearly in the origin of Australian Merinos to supply the fine wool previously exported from Spain and Germany to England. They are discernible as controlling the urge for greater fleece-weights by increasing the number of wool fibres per unit of skin surface which led to "developed" and "dense-woolled" sheep. They are now implicated in restricting "development" to reduce the ravages of blowflies which, as environmental factors, are little less predatory than the carnivora that worried Celtic Shorthorns in medieval England. To satisfy such changing requirements livestock populations must themselves be capable of change. Genetically this is most likely to be possible among heterozygous animals which, in the mass, result principally from crossbreeding, a process which requires heterogeneity.

CHAPTER II

ROBERT BAKEWELL

DEMANDS for animal products as food were clamorous in England early in the eighteenth century. The time was therefore opportune for somebody to evolve from the livestock at hand types suited to the new requirements. This was undertaken by Robert Bakewell, who was born in 1725 and succeeded to his father's farm in 1760. He bred sheep and cattle successfully and has been regarded as having pioneered improved methods of breeding.

The conditions of modification, however, were not implemented by him. The heterogeneity that characterized every kind of livestock in Great Britain had much earlier origin. All three principal environmental conditions, using the term in its widest sense, of enclosed land, of improved cultivation with suitable crops, and of outstandingly large demands for meat and milk, had developed and were developing beyond his control. A great number of anonymous breeders had prepared the way for him to select, from among their livestock, animals peculiarly suited to his own purposes. These were to develop animals characterized by early maturity, quick fattening, light bone, short legs, and a high percentage of carcass to live weight. Broadly, however, the history of British breeds of livestock began with Bakewell's era.

Prentice suggests that Bakewell adopted his policy of mating like to like from the early procedure for breeding English game cocks. Macdonald and Sinclair (1909) state that the first Tomkins who bred Herefords practised in-and-in breeding before Bakewell adopted the method. Thoroughbred horses also were "under way" before

Bakewell's time. However, his methods were followed by a number of disciples who modified other farm animals to satisfy the clamorous demand for foods. of animal origin. The constructive period, therefore, can be regarded as having been introduced by Bakewell's work with Leicester sheep and Longhorn cattle.

He and his successors, as improvers of the principal British breeds of beef cattle and mutton sheep, demonstrated their as.uteness at three points. They recognized the need for particular kinds of livestock; they appreciated, among the relatively unselected populations, the possibilities of individual animals; and they evolved methods by which the occurrence of their favoured types could be regularized.

The old-type Leicester sheep with which Bakewell began his work is reported as having been large; relatively coarse and slow-maturing. His ideal apparently was a smaller beast with neater appearance that fattened readily. He was secretive as to his methods, but Marshall, who was a contemporary of his, appears to have been in his confidence. He is quoted by Youatt (1837) as stating that, in improving the long-wools, Bakewell made

> no crosses with any alien breed. The improvement was effected by selecting individuals from kindred breeds—from the several breeds or varieties of long-wool sheep with which Mr Bakewell was surrounded on every side; by breeding in-and-in; by selection, solicitously seizing the superior accidental varieties produced; associating these varieties and still continuing to select, with judgment, the superior animals.

Bakewell restricted his objectives. For him carcass type was the first consideration. He valued it so highly that he neglected the fleece altogether. His principle was to reduce his fundamental requirements to the minimum. No individual pedigrees are available for his sheep, although it is known that he kept such records. Indeed, that he did so and that he was imitated by his disciples is probably his most outstanding contribution.

He began to work with cattle a good deal later than with sheep. Already the Longhorn breed had been improved to some extent, and Bakewell went to these improved herds for his foundation animals. One of his first purchases was a cow named Comely which was mated with a bull known as the Westmoreland bull because he was bought in that county. A Mr Fowler, who was a contemporary and client of Bakewell, followed his methods. He leased a bull "D" bred by Bakewell and with it bred a bull Shakespeare, said to have been a very successful sire. His pedigree is given (Fig. 1) with that of

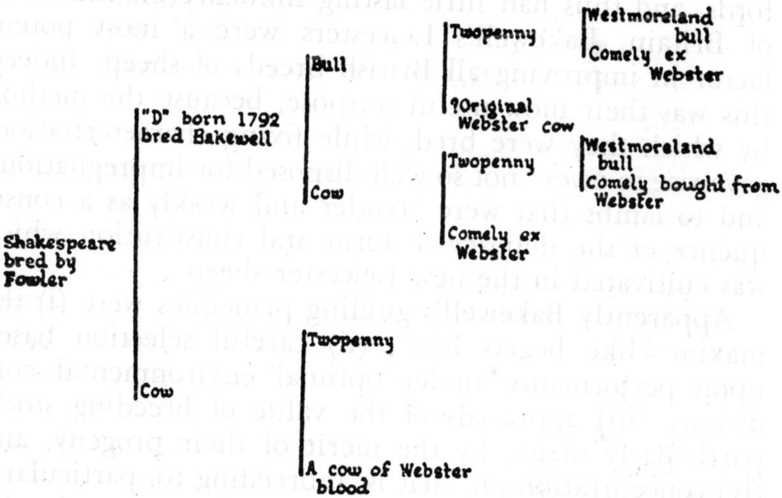

FIG. 1.

Pedigree of the bull Shakespeare to show the methods of Bakewell and Fowler. This pedigree is taken from Youatt's statement: " 'D' was got by a son of Twopenny out of a daughter and sister of the same bull, she being the produce of his own dam." Wilson (1909) shows Twopenny as the sire of "D".

the bull "D", because it is said of Shakespeare that "his horns apart, he had every point of a Holderness or a Teeswater bull. Could his horns have been changed, he would have passed in Yorkshire as an ordinary bull of either of those breeds."

The end point of Bakewell's inbreeding was Comely, that of Fowler's was Twopenny. Twopenny was related equally to the Westmoreland bull and to Comely. The county of Westmoreland adjoined that of Durham, the home of Teeswater and Holderness cattle. It is not surprising, therefore, that Shakespeare, "his horns apart, had every point of a Holderness or a Teeswater bull". The "accidental" variety of that day would now be expected from the method. As a result of the continual inbreeding, parental types were segregating.

It is known that Bakewell fed his livestock generously and that he leased sires, both rams and bulls. Although his Longhorns were displaced by Shorthorns and Herefords, and thus had little lasting influence on the cattle of Britain, Bakewell's Leicesters were a most potent factor in improving all British breeds of sheep. Indeed this was their most useful purpose, because the method by which they were bred, while fixing characterization, gave rise to ewes "not so well disposed for impregnation" and to lambs that were "tender and weakly as a consequence of the delicacy of form and constitution which was cultivated in the new Leicester sheep".

Apparently Bakewell's guiding principles were (i) the maxim "like begets like", (ii) careful selection based upon performance under optimal environmental conditions, (iii) appraisals of the value of breeding stock, particularly males, by the merit of their progeny, and (iv) concentration on, that is, inbreeding to, particularly favoured animals. This involved pedigree records. All of these can be deduced from his work and on these grounds he has been referred to as "the father of animal breeding". It is probable, however, that his accession to the title is fortuitous. It is easy, when prepared by biological knowledge, to read into such records more than was ever conceived by the person who established them.

Bakewell, began breeding sheep about 1750, he bought the Westmoreland bull in 1760 and he died in 1795. Before that date there was insufficient knowledge for him to have had even a clue to causes underlying the processes he employed. Observational investigation of anatomy, physiology, and histology was begun before the close of the seventeenth century. In each case, however, work had been restricted to a few animals. Cuvier (1769-1832) and Bichat (1771-1801) extended this knowledge by their respective studies of comparative anatomy and the structure of animal tissues. However well informed Bakewell may have been on these matters, the broad conceptions of biology necessary for his work to have been other than empirical were not put forward until dates subsequent to 1850. In Bakewell's day, each species and each variant was regarded as an independent creation and offspring of animals were considered to be a blend of the parental types.

His unorthodox method of continued in-and-in breeding may have been inspired by the Bible. The Book of Leviticus states, as an injunction, "Thou shalt not let thy cattle gender with a diverse kind; thou shalt not sow thy field with mingled seed; neither shall a garment mingled of linen and wool come upon thee."

There could be no stronger instruction to maintain the isolation and purity of breeds than this trebly reinforced command. This instruction from a book to which he must have had ready access; or his own observations and knowledge of the procedure of others, for example the eldest Tomkins; or an idea, so common to breeders, that no other animals are fit to mate with their own—any of these may have been his inspiration to pursue his policy of in-and-in breeding. It is impossible to determine which it was, but there can be no doubt that he appreciated the fact that by this method he isolated his animals from others. Offspring of his flocks and herds were descendants of a few isolated animals. By their relationship to one another they became similar to each other and to their ancestors, but dissimilar to other animals of the same breed but having different relations.

Bakewell subscribed to this principle of isolation, but whether he did so for biological or for commercial reasons must remain open to doubt. He did, however, found the Dishley Society with thirteen rules to which every member had to subscribe. The first was, "No member shall hire or use a ram not belonging to Mr Bakewell or to one of the members of the Society."

If the Dishley Leicesters were all that could be desired this was biologically sound, and, in any case, it was good business. The interesting point is that by this and the remaining rules of the Society a group of related animals, scattered throughout a number of flocks, was isolated as a breeding commonwealth. This conception has become the credo of breeders of "pure-bred" livestock, as witness the number of flockbooks and herdbooks published by as many breed societies. The first of these was prepared at the instigation of the Colling brothers, who were disciples of Bakewell. It was published by Coates for Shorthorns in 1822 and has since been followed widely as a precedent.

CHAPTER III

SHORTHORNS*

THE Shorthorn breed of cattle, although not unique in this respect, had a cosmopolitan ancestry. Derived from native British, imported Roman, and then Dutch cattle, it became as typically English as Westminster Abbey. Its centre then swung to Scotland, where it developed specific characters, and from there northern types successfully invaded the original home of the breed. During the 115 years of their history that are briefly reviewed, these cattle were transferred to most parts of the world and there set production standards for both beef and milk. The first of the herdbooks was prepared for the breed and its interesting development, as traced in subsequent volumes, has attracted historians and biologists to record and examine the methods of its construction.

There was a long anonymous period that preceded recorded history of Shorthorns. During that period a high standard was reached, but apparently no specific effort had been made to unify their characteristics. This began with the work of the brothers Colling who were contemporaries and pupils of Bakewell. From 1783 onward, the breed history of Shorthorns can be described as a series of tableaux with foregrounds and backgrounds, production being controlled by the ghost of Bakewell, the Leicestershire farmer. This is indicated in Fig. 2.

The foreground has been occupied by men who, each in his own time, made a major contribution to the construction of the breed. The limelight of history illumi-

---

* The author is considerably indebted for information in this section to the *History of Shorthorn Cattle* by Macdonald and Sinclair. Vinton and Company, London.

nates their work, while countless other figures of owners and herds move dimly in the background. From these herds the central figures drew foundation stock. With them they developed closely related animals for sale or lease and returned to herds in the backstage groups for rejuvenating outcrosses.

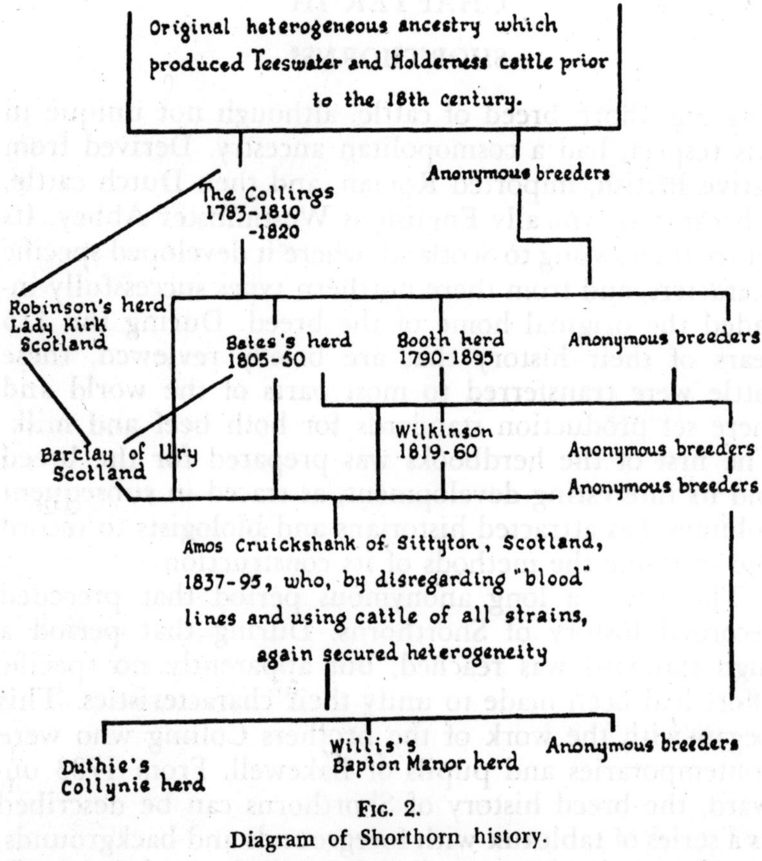

FIG. 2.
Diagram of Shorthorn history.

THE COLLING BROTHERS

Whereas Bakewell was apparently able to purchase foundation animals that suited him, for example, the cow Comely, the Colling brothers had to amalgamate contrasting types. They took advantage of the system then in vogue whereby cows were received for service by outstanding bulls of other owners. In pedigrees of cattle purchased by the Collings and of cattle of their early breeding, whether by their own or by outside bulls, there is distinct evidence of compensatory matings, that is, matings between dissimilar animals. Youatt states that Colling found the Holderness and Teeswater cattle "like all other extravagantly large cattle, frequently of loose make and disproportion". The bull Hubback was at that time a calf, running with his dam on the roadside, and owned by a person in indigent circumstances. Both the cow and her calf were peculiarly capable of ready fattening. It is stated by Youatt that when they were put on good food, the cow fattened and did not breed again while her son

> . . . having the same feeding propensity in a high degree, was useful as a bull during a very short period. The quality of his flesh, hide and hair are supposed to have been seldom equalled and, as he was smaller than the Teeswater cattle, he was eminently calculated to forward Mr Colling's views.

Colling reserved this bull for use among his own cows, the result being, as stated by Youatt in 1874, "we have no superior Shorthorns which do not claim descent, nearly or remotely, from Hubback".

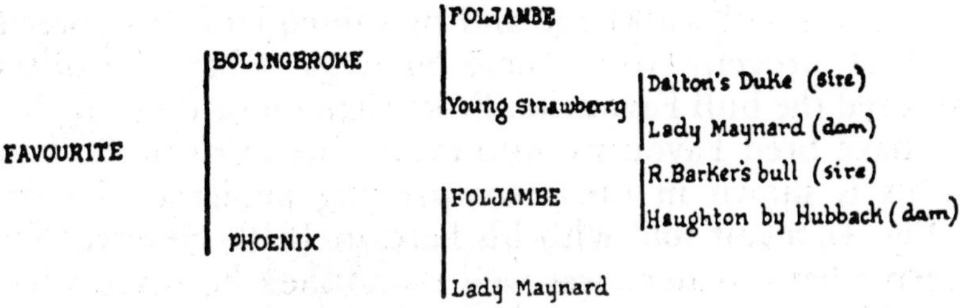

FIG. 3.

Pedigree of Favourite bred by Charles Colling. Animals with names in large type were bred by Colling.

Fig. 3 and the text show compensatory matings in the pedigree of Favourite, the bull for which Charles Colling is most famous. The names in block print are those of cattle bred by Charles Colling. He purchased Young Strawberry, Lady Maynard, and Haughton. R. Barker's bull was one to which he sent Haughton for service. Descriptions of the cattle are taken principally from Sinclair's *History of Shorthorn Cattle.**

Hubback:	A small neat early-fleshing bull.
Haughton:	A neat fine cow the reverse of R. Barker's bull.
Richard Barker's bull:	A large, coarse, wire-haired beast with a large head, dark horns and a black nose. His dam, a big coarse cow.
Lady Maynard:	The beautiful Lady Maynard.
Dalton's Duke:	Brindle and coarse.
Foljambe:	A useful thick beast, good handle and dark face.
Young Strawberry:	No description available.
Phoenix:	Handling particularly soft and mellow; in appearance the reverse of her dam, the beautiful Lady Maynard.
Bolingbroke:	Remarkable inclination to fatten, handsome hindquarters, fine head and muzzle.
Favourite:	Large massive animal of good constitution, had remarkably good loins and long level hindquarters.

By a series of matings in which coarse was mated with fine and small with large and by subsequent close breeding of progeny from these matings, Charles Colling secured the bull Favourite. From then onward he appears to have bred Favourite into every line of cattle he had. This is shown in the accompanying pedigree of Cecil (Fig. 4), a calf sold with his herd in 1810. Nevertheless, even when his herd was well established he made what, for him, must have been a violent outcross. He introduced the "blood" of a red Galloway cow into his herd as an "alloy". The caution of his method in doing so is shown in Fig. 5. It is worthy of study because every breeder who has developed a strain of animals "closed to outside blood" has at some time approached this problem with trepidation.

* Sinclair in his *History of Shorthorn Cattle*, gives his authorities for these and other descriptions and pedigrees used extensively in this review.

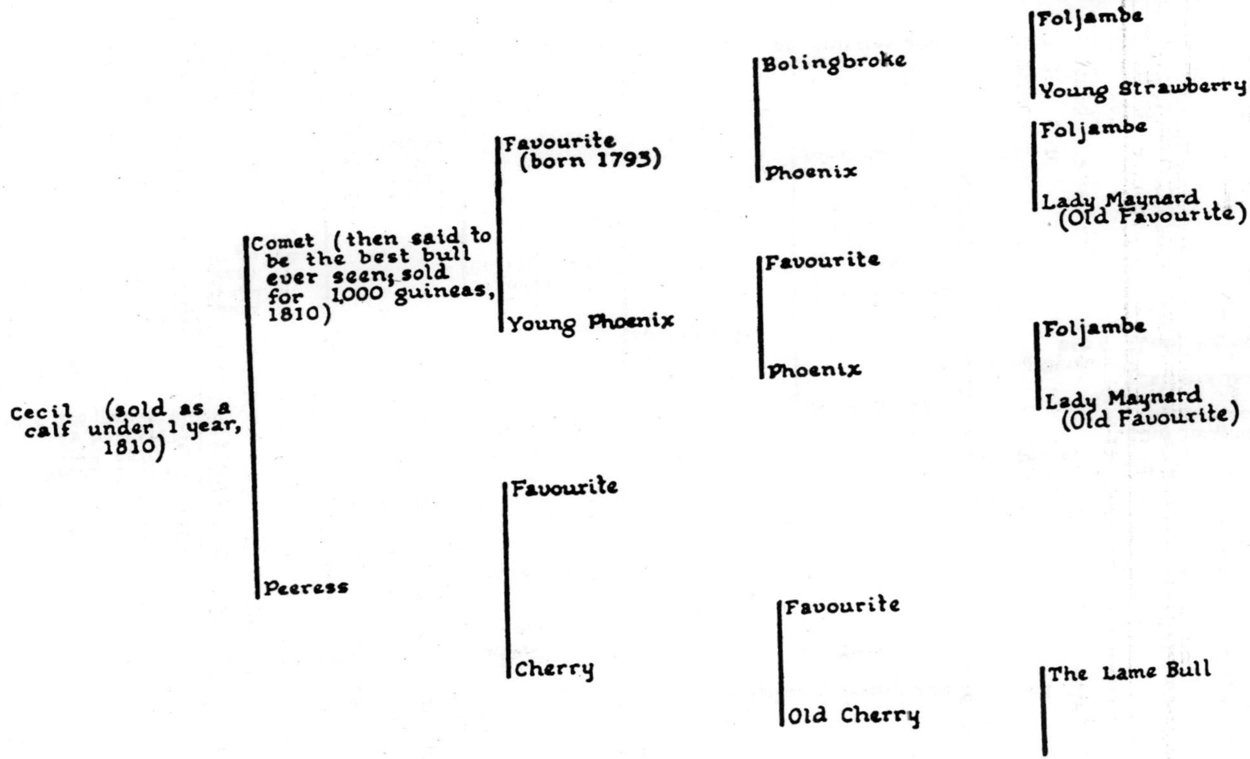

FIG. 4.
To show how Colling inbred to Favourite.

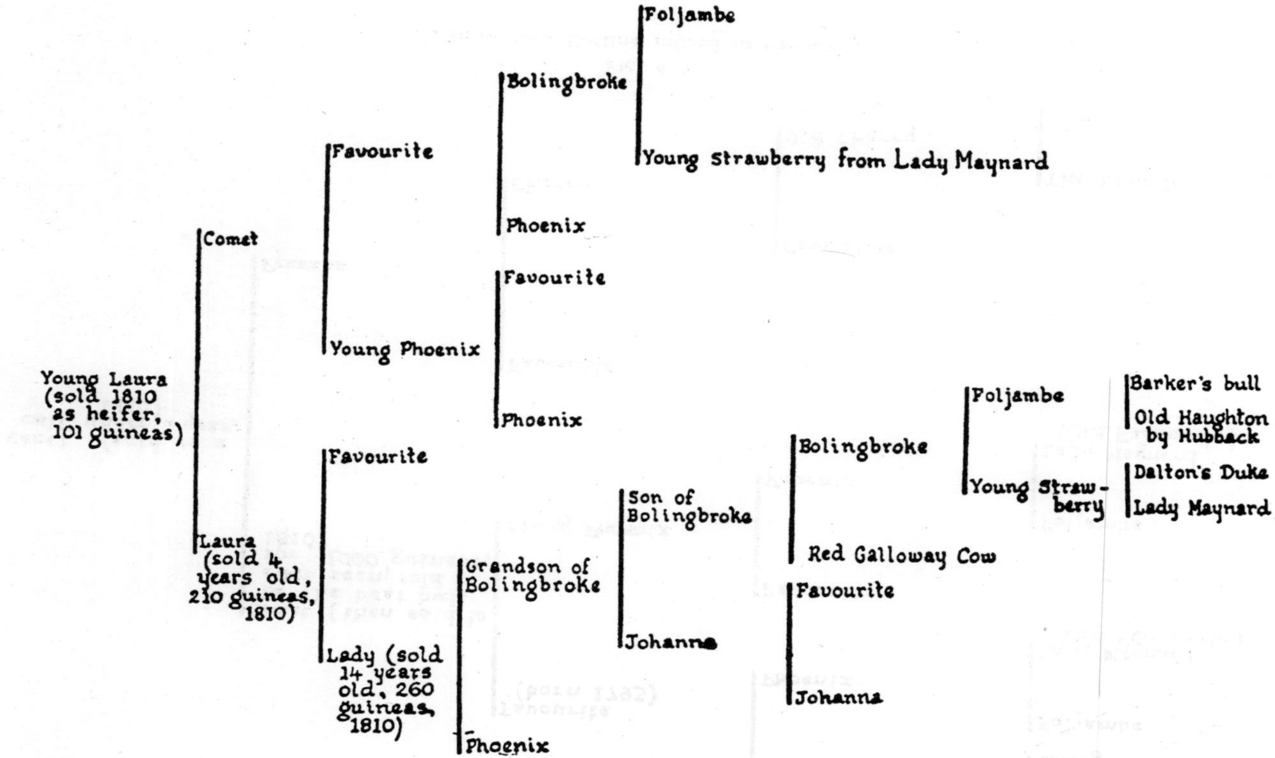

FIG. 5.

Showing how Colling introduced a red Galloway cow to make an outcross and an "alloy" in his herd.

THE BOOTH FAMILY

The Booth herd is of special interest because it was a "long-distance" herd. The family maintained it for three generations of Booths. Their methods of periodic rejuvenation are particularly interesting because the same problem frequently arises in closely bred herds of long standing, especially if the breeding group is relatively small.

The first member of the family, Thomas Booth, secured his foundation animals some time before 1790, wherever he found suitable stock among the original Teeswater cattle. With them he mated Colling bulls. This initiated a series of matings in which an inbred line of bulls, with relatively fixed characters and always from the same herd, was mated with cows at first only very distantly related to the line. It was in effect a period of grading-up by using top crosses of Colling's bulls. When the Booth type was established by the method and the selection that accompanied it, home-bred bulls were used. From this basis five members of the Booth family bred Shorthorns prior to 1908, when Sinclair stated that a member of the family was still breeding them. The early history of the venture is associated with the herds of the grandparent, Thomas Booth, of his two sons, John and Richard, and with those of his two grandsons, T.C. and J.B.

The founder used four principal bulls from Colling. They were, (i) Twin Brother to Ben by Punch, dam by Foljambe; (ii) Suwarrow by Styford from a cow by Favourite (Styford was by Favourite from a sister in blood to Twin Brother to Ben); (iii) Brother to Young Strawberry by Dalton's Duke from Lady Maynard (Young Strawberry was paternal grandam of Favourite); (iv) Albion by Comet, a son of Favourite from a daughter of Favourite, dam Beauty, by a son of Favourite. Reference to the pedigree of Cecil (Fig. 4) will disclose how many of these cattle were bred; how Thomas Booth used

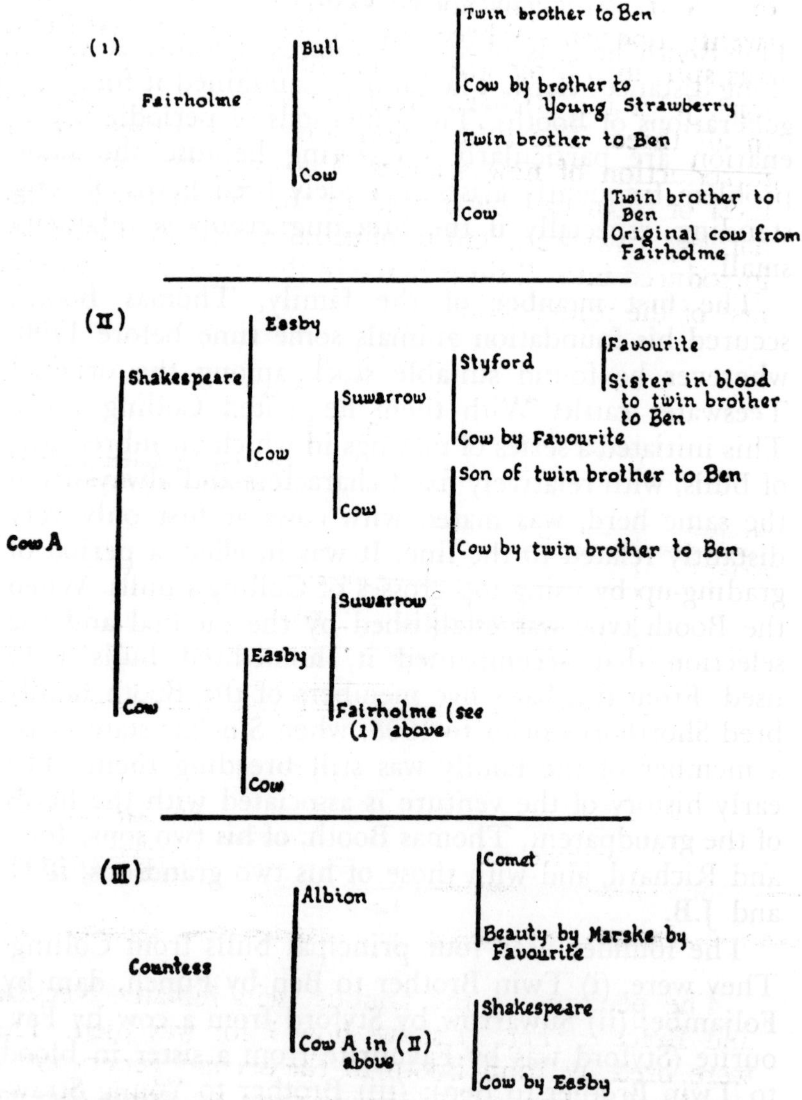

FIG. 6.
Pedigrees of cattle from Fairholme to Countess, showing matings in the earliest Booth herd.

these four bulls consecutively is shown in the pedigree of the cow Countess (Fig. 6).

The next bull used was Pilot. He was bred by Robert Colling and was by one or other of two bulls Wellington or Major, father and son respectively. In 1819, after twenty-nine years of breeding, the herd of Thomas Booth was split into three groups. He held one, while his two sons, John and Richard, held the others. Pilot was used in all three. The subdivision was accompanied by the introduction of new females. However, the parents of most of them had been bred in the Booth herd. Alternatively, cows were sent to outside bulls. New blood was introduced into all three herds in this manner. It was the first of the rejuvenations.

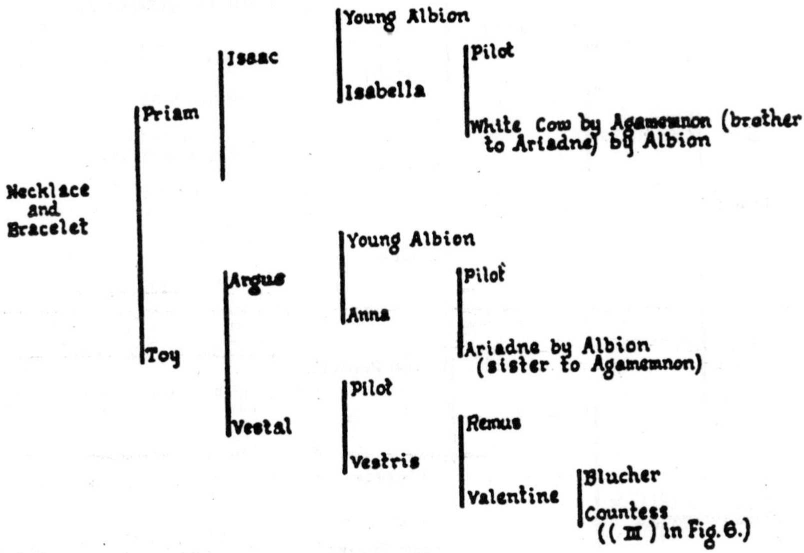

FIG. 7.
Use of Pilot in the Booth herd.

The pedigree of the famous twin females Necklace and Bracelet (Fig. 7) shows how Pilot was used. They were bred by John Booth in the second generation of the family. The pedigree continues on from the cow Countess shown in Fig. 6 (iii).

Although the pedigree of Necklace and Bracelet shows less immediate inbreeding than some that have appeared earlier, it should not be thought that the process had been discontinued or reduced. The bull Jeweller, bred about this time but sold, and reported to have been a successful sire, was considerably inbred. Both of his parents were by Leonard. His sire was from Bracelet and his dam was from her twin sister Necklace.

Thomas Booth died in 1835 after forty-five years of Shorthorn breeding. Two new bulls were used relatively soon after his death. Sinclair considers that by their introduction the son Richard "renovated his father's herd". This took place sixteen years after the earlier rejuvena-

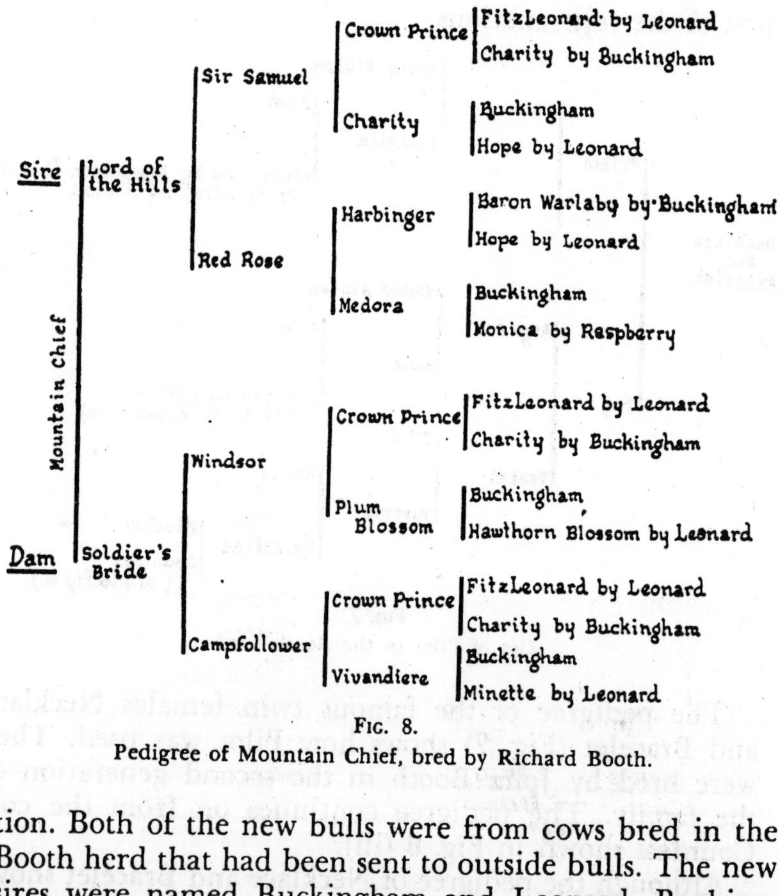

FIG. 8.
Pedigree of Mountain Chief, bred by Richard Booth.

tion. Both of the new bulls were from cows bred in the Booth herd that had been sent to outside bulls. The new sires were named Buckingham and Leonard. Buckingham was by Mussulman (brought out to Victoria by the Boldens about 1840) from Bracelet, while Leonard was by Lord Lieutenant from White Strawberry. This cow,

White Strawberry, was by Rockingham by Isaac, who was by Albion from Isabella by Pilot.

Fig. 8 shows that Richard Booth used these two bulls by methods similar to those that had preceded the two rejuvenations of the herd. When Richard Booth died in 1864 his herd was taken over by a nephew, T. C. Booth, who introduced new blood, both male and female. This was about thirty years after the more recent "renovation". The time intervals between them were approximately 29, 16, and 30 years in a total of about 75 years of herd life. Lush (1937) states an average cattle generation to be approximately 4·5 years. These intervals therefore represent something in the order of 6, 4, 7, and 17 cow generations respectively.

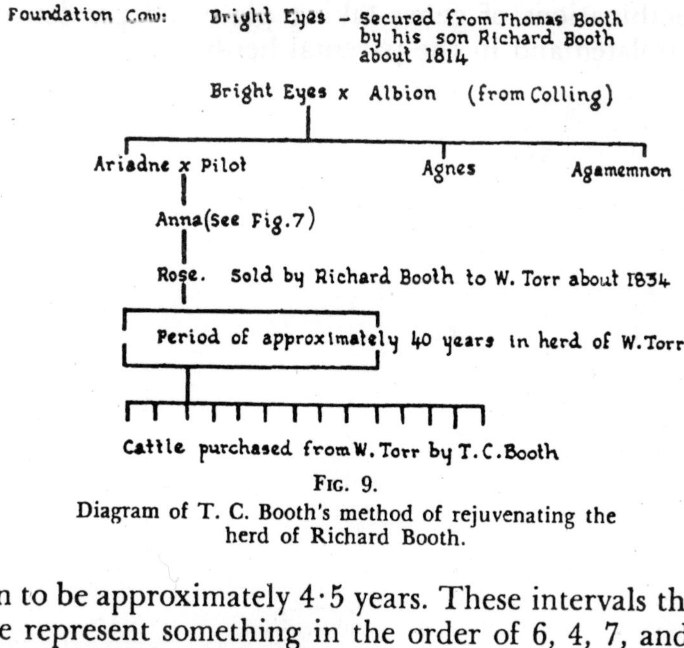

FIG. 9.
Diagram of T. C. Booth's method of rejuvenating the herd of Richard Booth.

The method of rejuvenation used by T. C. Booth, although different from those previously employed, was, in principle, typical of Booth procedure. He purchased from W. Torr at Alsby twelve head of females that were descended, in that herd, from cattle purchased from Booths some forty years earlier. This is shown diagrammatically in Fig. 9. There was no departure from Booth "blood lines" because all the bulls used at Alsby had been bred either by the Booths or by Torr from Booth-bred stock. This variation of the Booth method was used later by owners of Bates-bred cattle when, under similar conditions, Duchess cattle were returned from America to enter particular herds in England. It has been referred to as a "climatic outcross": the animals have similar pedigrees but have been bred elsewhere.

Genetical dissimilarity of such cattle, and so their suitability as outcrosses, arises from differences originating in the genotypes withdrawn as samples of those comprising the original herd. Differences would arise also from recombinations of genes taking place independently in the isolated and in the parental herds.

CHAPTER IV

SHORTHORNS (CONTINUED)

THOMAS BATES OF KIRKLEVINGTON

THOMAS BATES was a contemporary and keen rival of Richard Booth. Each developed a distinctive type. The slogan was "Booth for the butcher, Bates for the pail". Both herds had many partisans, but there was an increasing number of "backstage" breeders who, remaining aloof from the controversy as to which was the better, had "a bit of both". Some time before the dispersal of Colling's herd, Bates bought a cow named Duchess. At the sale, in 1810, he bought her heifer by Comet. From these two females, but particularly from the heifer, which he called Duchess 1st, he developed the Duchess family of Shorthorns.

Broadly, Bates's methods were based upon a constant degree of close mating, restricted to particular blood lines, all of which traced to the same ancestors. He was not particular with regard to the number of intervening individuals. From 1812, when the Duchess heifer by Comet bore her first calf, until 1849, when he died, Bates bred 63 cows in this Duchess family. They were named Duchess 2nd to Duchess 64th. Forty-five males are recorded as dropped by Duchess cows, 29 of which were named and recorded. When his herd was dispersed, eight females and six bulls of the Duchess family were sold, together with 44 cows and ten bulls from five other "families".

The Comet Duchess was first mated with a bull named Ketton, by Favourite from the original Duchess cow, mother of the Comet heifer. The second bull, Ketton 2nd, was by Ketton out of a granddaughter of Favourite. The third bull used was Ketton 3rd by Ketton 2nd out of Duchess 3rd. She was by Ketton from the Comet Duchess. Marske, the fourth bull used, was by Favourite from a cow by Favourite out of a daughter of Favourite. The bull Cleveland followed. He was by Ketton 3rd from the Comet Duchess. Young Marske, another bull used, was by Marske from Duchess 4th. She was by Ketton 2nd from the Comet Duchess. The Earl was by Duke, full brother to the Comet Duchess, from Duchess 3rd. Second Earl was by the Earl from Duchess 3rd. Third Earl was by the Earl from Duchess 8th by Marske.

At this stage Bates evidently considered that he should call a halt and introduce new blood. The calf most recently born at that time was by 3rd Earl from Duchess 11th which was by Young Marske out of 5th Duchess. The calf had a coefficient of inbreeding (Wright* 1921) of 44 per cent. How intense this is can be seen from Fig. 10. The offspring X in that figure has a coefficient of inbreeding of 43·75 per cent after four generations, all by the same sire. Bates retained his whole family of Duchess cattle at about this level of inbreeding.

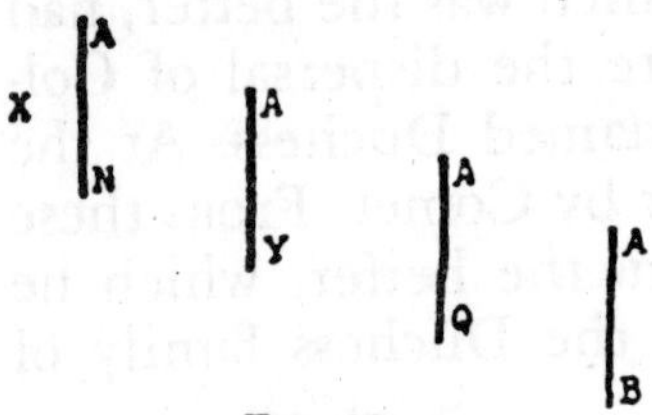

FIG. 10.

Diagram of a simple pedigree to show inbreeding which has a Wright's coefficient of 43·75 per cent.

Bates introduced new blood by securing the bull Hubback 2nd. He was by Earl from Red Rose. The bull Earl already had been used extensively in the herd, but Red Rose had no Duchess blood. However, she had similar ancestry. Favourite was her sire, her great-grandsire and her great-great-grandsire in three lines of her descent.

When Bates had fifty cows in his herd, all by Hubback 2nd, he again introduced new blood by purchasing the Princess bull, Belvedere. This bull was intensely inbred, resulting from mating a full brother and sister to be his parents. He had no Duchess blood, but his relationship

* Wright's coefficients of inbreeding and relationship are stated in the appendix.

to Favourite was about 65 per cent. Wright (1923) has shown that their common descent from Favourite established a relationship of over 40 per cent between Belvedere and the two earliest Duchess cows with whom he was mated. For comparison, the relationship of father to son is 50 per cent.

Belvedere was followed by his sons, bred as is shown in Fig. 11. Both bulls, Duke of Northumberland and 2nd Duke of Northumberland, were used on Duchess females. Pedigree of the cow Duchess 63rd, the last but one of the Duchesses bred by Bates, is presented in Fig. 12 to show the general method which can be followed back

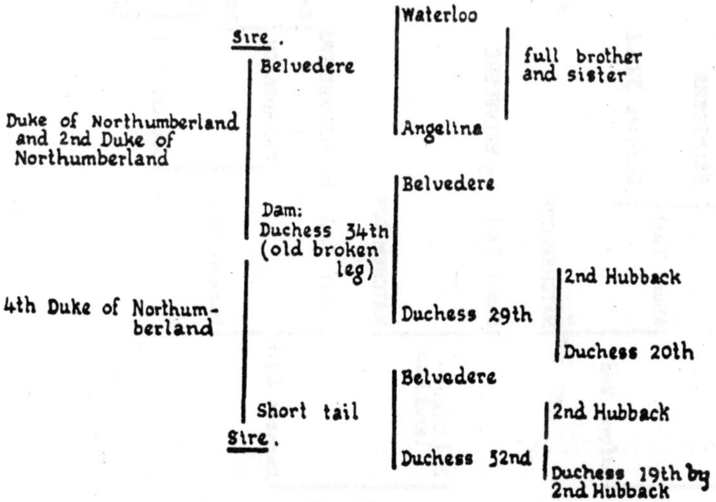

FIG. 11.
Use of Belvedere by Bates.

to the first or Comet Duchess. Unfortunately not all the animals introduced by Bates occur in this pedigree, but those that do are in block type.

Wright, in his study, *The Duchess Shorthorns* (1923), concluded that from the beginning to the end of his breeding career, a period of about forty years, Bates kept the degree of inbreeding and of relationship within his herd from rising above a certain level by introducing dashes of outside blood. He maintained an average degree

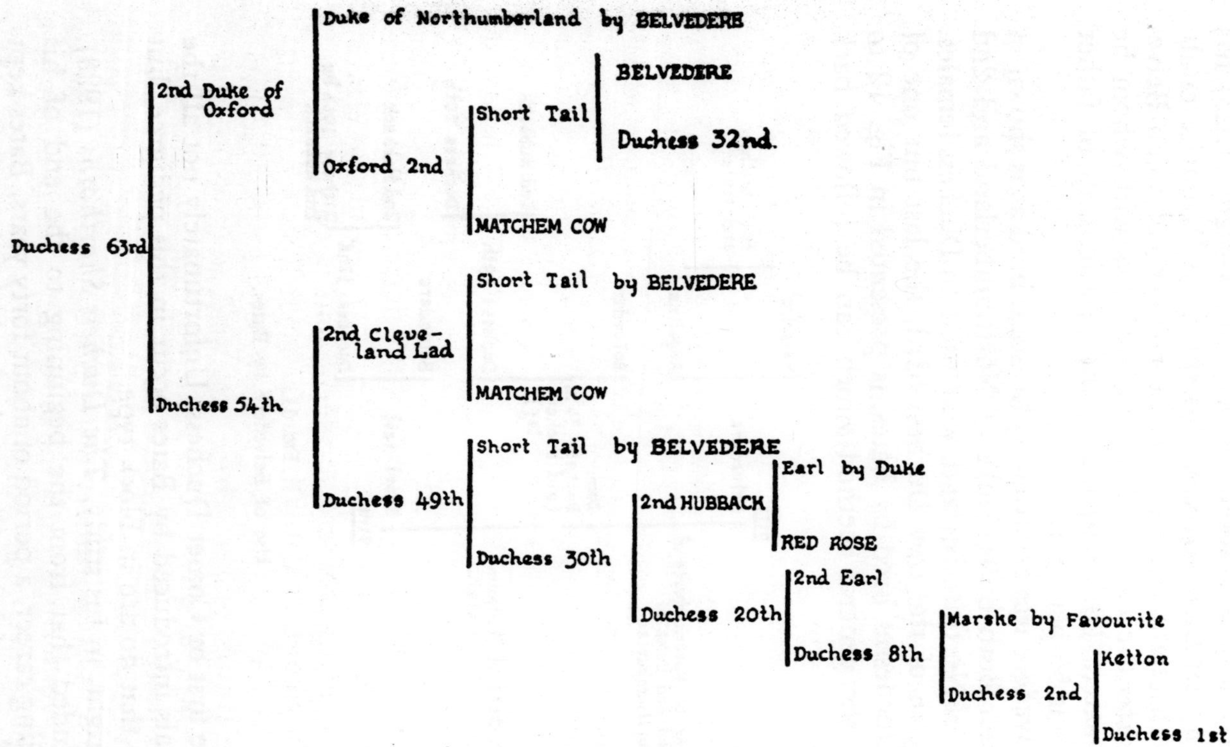

FIG. 12.

Breeding method of Bates as shown in pedigree of Duchess 63rd. Names of introduced stock are in block type.

of relationship between the animals he mated calculated by Wright to be distinctly higher, namely by 10 per cent, than that of 50 per cent between an ordinary brother and sister.

The level of inbreeding was about 40 per cent. It is approximately equivalent to two generations of straight-out brother-sister matings. The degree of relationship of all the cows to Favourite was held at between 57 and 69 per cent, despite the fact that he was the sire of only two of the early Duchesses bred by Colling. This resulted from the high degree of relationship to the bull Favourite common to all early Shorthorns.

While Bates maintained these relationships he constantly selected for certain types of cattle. Indeed, upon at least one occasion he sold out a number of lines. Wright considers that, had Bates chosen a type fixed by a higher degree of inbreeding, he would probably have been unable to practise the necessary selection. Had he maintained less inbreeding, it is also probable that "simultaneous variations of many characters would have been more than he could contend with". Wright considers that Bates maintained "a happy medium between these two probabilities, and so he retained the plasticity of his stock".

Bates's methods, however, appear to have had a serious effect on the fertility of his cattle. There are many records of bulls being temporarily sterile and of barren or shy-breeding cows. Eventually the Duchess line of cattle disappeared. Among the last of their kind were some bred in the United States. These were sold at the Mills Sale, New York, in 1873. The value placed upon so-called "pure descent" and upon "purity of pedigree" overshadowed so greatly all other considerations that this sale resulted in extraordinary prices for the last of this lowly fertile strain. Apparently the fewer they became the more highly was their rarity valued. Actually, two cows bought at the sale for approximately four and five thousand pounds, respectively, have no recorded progeny. High prices were paid for all of the two bulls and eleven cows of the Duchess family. English buyers who desired them for their own herds bought one bull and five cows for 29,420 guineas, average 4903 guineas, while American buyers, who wished to retain the cattle, purchased the remainder for 22,260 guineas, average 3180 guineas. At the dispersal of Bates's own herd in 1850, there were six bulls and eight females of the Duchess family. They were bought at auction for an average of £116 5s. and a total of £1627 10s. in an aggregate of £4558 1s. for the whole herd of 68 animals.

SCOTTISH SHORTHORNS

Amos Cruickshank, in partnership with his brother, bought his first Shorthorn in 1837 and established a herd at Sittyton, near Aberdeen, Scotland. His objective was a rugged, deep-fleshed animal of great vigour. "He assiduously sought those with outstanding constitution and disregarded current opinion with regard to the affinities of particular 'blood lines'."

Most of his early sales of bulls were made, not to fashionable breeders, but to farmers who relied largely upon locally-grown straw and turnips to fatten crossbred steers. They were not impressed by pedigrees, but appreciated an ability to lay on flesh with a minimum of keep. With similar ideals Amos Cruickshank toured the studs and attended sales. Always he judged the individual for what it was rather than by the fashion value of its pedigree.

This independent northern outlook was strange to breeders who had followed precedent. It was adopted by Cruickshank when the craze for "pure pedigree" was gathering strength to reach its highest point at W. Torr's sale, where eighty-three head brought £43,000. Cruickshank's attitude in this matter may have been dictated by his financial resources, because his brother bought "pure line" cattle as opportunity offered. However, genetically it was one of the soundest possible moves. It revived part of the original heterogeneity, much of which had been lost to the breed, by too close adherence to particular blood lines with consequent in-and-in breeding. It was even more iconoclastic in this breed, and as critical as the resolution of colour differences among breeders of Herefords. Each served a similar purpose.

Cruickshank's independence of thought controlled the early matings at Sittyton. He continued to buy vigorous cattle, mating them as individuals rather than as members of particular families or "blood" groups. In a general way he favoured cattle of Booth descent and avoided those of Bates origin. He bought a number of cattle and developed a herd of some two hundred head on an area of approximately one thousand acres. He thus operated on a larger scale than most of the other breeders. This provided him with a relatively wide range for selection and enabled him to keep in use a number of herd bulls.

Matings almost at random with regard to pedigree continued for approximately twenty-one years, nearly half the life of the herd. Then Cruickshank, in search of a young bull, visited the little-known English herd of a Mr Wilkinson. He made no immediate purchase but continued his tour of the studs. Later, in 1858, he arranged by correspondence to buy from Wilkinson for thirty guineas an inbred eight-year-old bull named Lancaster Comet.

Wilkinson would have remained one of the anonymous breeders of this review if Cruickshank had not acknowledged his great indebtedness to this bull. Wilkinson had bred closely for twelve years with a bull named Will Honeycomb, bought by him in 1836. This bull was described as

> . . . very compact, on short legs, of great substance, with good head and neck, beautiful shoulders, breast very wide and nearly down to the knees, well fitted up at the back of the shoulders and along his sides, wide back with ribs well sprung and deep, splendid thighs well let down, hips well covered and not too prominent, quality first-rate with beautiful hair, hindquarters rather short but tail well set, colour, rich dark roan.

This description would fit many of the best present-day Scotch Shorthorns, and Amos Cruickshank is quoted by

Sinclair as saying that this bull was their principal ancestor.

When Will Honeycomb died Wilkinson continued to breed back to him, using his sons and their sons from cows purchased to introduce new blood. One of the double grandsons, by a son from a daughter of Will Honeycomb, was Lancaster Comet. This bull, in the ownership of Cruickshank, became the sire of Champion of England (Fig. 13). The paternal grandam of Lancaster Comet was Queen of Trumps, bought by Wilkinson about the same time as he bought Will Honeycomb. She

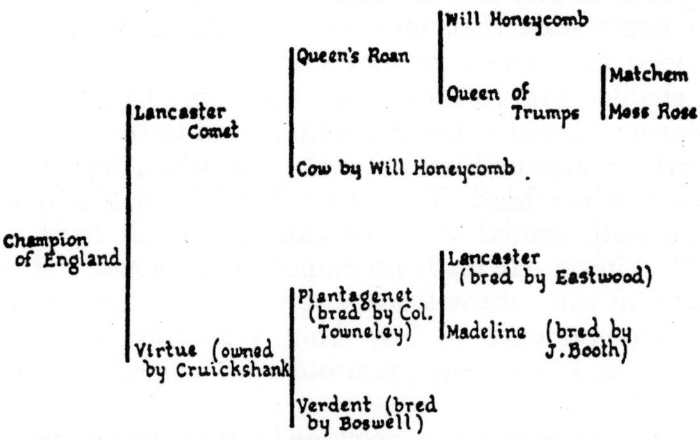

Fig. 13.

Pedigree of Lancaster Comet and Champion of England.

was by Matchem from Moss Rose, which was a closely bred animal with much the same breeding as the bull Belvedere used extensively by Bates. Queen of Trumps was described as being of small size but perfect in form. By Will Honeycomb she bore Queen's Roan, the sire of Lancaster Comet.

Apparently Amos Cruickshank did not exclude the minor point of horn type from his list of requirements for a bull. When Lancaster Comet was untrucked both bystanders and Cruickshank likened his horns to those of a Highland beast, and the bull was put on a remote farm to hide his atypical spread. When the mating season arrived he was turned out with some cows that were difficult to get in calf. He was left out too late in the autumn, developed rheumatism and had to be slaughtered in that year. Little more thought was given to the loss of a thirty-guinea purchase with atypical horns and an unduly large head.

Under these conditions Lancaster Comet sired about twelve calves. Little value was placed upon them, for most of them were sold. However, one was Champion of England, founder of the strain of Scotch Shorthorns associated throughout the cattle world with the names of Cruickshank of Sittyton and Duthie of Collynie. He was the senior sire, intensively used in a great herd, and pro-

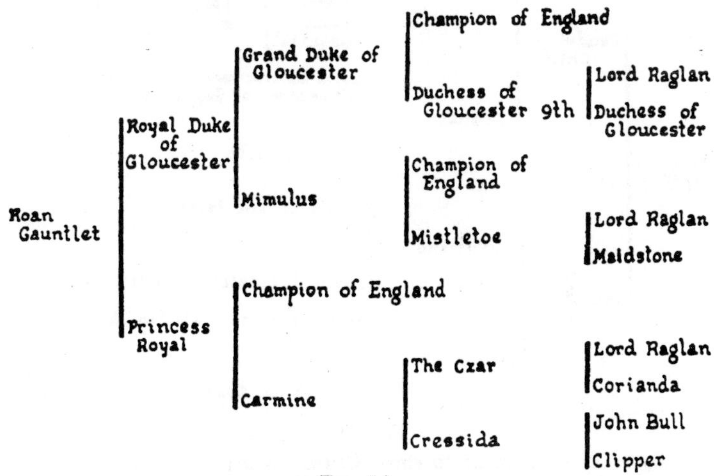

FIG. 14.
Pedigree of Roan Gauntlet to show Cruickshank's use of Champion of England.

genitor of bulls sold for thousands of guineas to carry the strain to other countries. So does chance operate; so can a minor point influence the judgment of an otherwise realistic breeder.

The second period, of approximately thirty years, in the Sittyton herd began with the discovery of Champion of England. Although Cruickshank, until 1874, continued to buy other bulls, he mated many of his cattle back to Champion of England by using his sons and grandsons as is shown in the pedigree of Roan Gauntlet (Fig. 14), a bull that he also used extensively. This was in the last fifteen years of the stud, in which only home-bred bulls were used. Champion of England himself was used for ten consecutive seasons. Fourteen sons of his were tried in the herd, but none was considered the equal of his sire. Sinclair, who was contemporary with Cruickshank, stated that the inbreeding within the herd in its

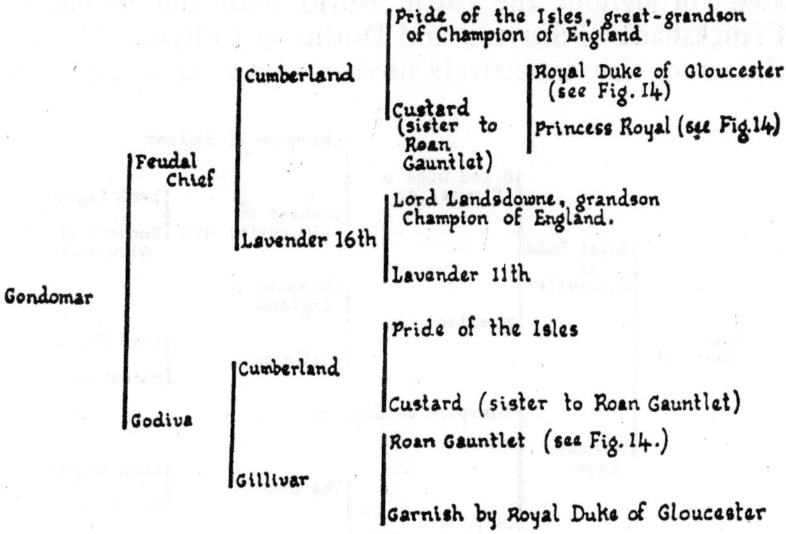

FIG. 15.
Pedigree of Gondomar to show Cruickshank's breeding method in the late years of Sittyton.

latest years was not because Cruickshank favoured the policy but because he regarded himself as being too old to initiate again the experimental matings necessary to overcome the difficulty. Roan Gauntlet was considerably inbred. He had been sold as a calf and used in an outside herd for three years prior to his recovery by Cruickshank in 1878. Although he was the most highly valued son of his father, it will be seen from the pedigree of Gondomar (Fig. 15) that Cruickshank even then did not breed as closely to him as either Booth or Bates probably would have done.

In 1889 the full Sittyton herd was sold by private treaty to go to the Argentine as a unit. Shipping and other difficulties prevented the purchasers from achieving their objective. It was thus that William Duthie of Collynie, Scotland, and J. Deane Willis of Wiltshire, England, were enabled to secure females from the herd. Collynie had used Sittyton bulls for a number of years, so the purchase under these conditions of over thirty females and several bulls from the Sittyton herd only amplified pre-existing blood lines. The Bapton Manor herd of J. Deane Willis was founded on English types from Bates and Booth lines. About 1890 it was reformed on the Sittyton purchase. These two herds carried on breeding the Scotch type of Shorthorn, the most recent modification of the original material from which Colling, Booth, Bates and Cruickshank successively evolved distinctive strains all highly regarded in their day.

CHAPTER V

ABERDEEN-ANGUS*

DARK-COLOURED polled cattle have been in Scotland at least since 300 B.C. There was only a small infiltration of Dutch or other cattle, and the origin of the polled cattle now known as Aberdeen-Angus was relatively homogeneous. In an unimproved state they had several shades of colour and were divided into those of Buchan and those of Angus, but had their origin in indigenous stock. Some were horned and some were polled but horned or polled, they had been valued highly in Scotland for many years before Hugh Watson in 1808 became the tenant of Keillor. It was he who first undertook their improvement.

Watson visited the English Shorthorn country in 1815 or 1816 and there came under the influence of Booth, Maynard, Wetherell, Torr, and other breeders. When he returned to Scotland he implemented their methods in the development of local cattle. It is stated that he mated the best with the best regardless of blood affinities; that he practised in-and-in breeding to a considerable extent but did not deliberately breed within "families"; and that he was forced to use his own bulls because they were better than any he could find elsewhere.

KEILLOR JOCKS

The first of three interesting points in the history of Aberdeen-Angus cattle is centred in the Keillor Jocks. This line of bulls was the most valuable contribution made by Watson to the development of the breed. The

---

* The author is considerably indebted for information in this section to the *History of Aberdeen-Angus Cattle* by Macdonald and Sinclair. Vinton and Company, London.

original Jock was Tarnty Jock, bought by Watson at a market in 1808. The second Jock was by him from a cow bred by Watson's father. From then on, each Jock was the son of the bull of that name which had preceded him and, in some cases, from a cow closely related to his sire. The method was followed until the fifth bull was born. He was known as Old Jock and became Watson's principal sire. The pedigrees of Old Jock and of two of his sons

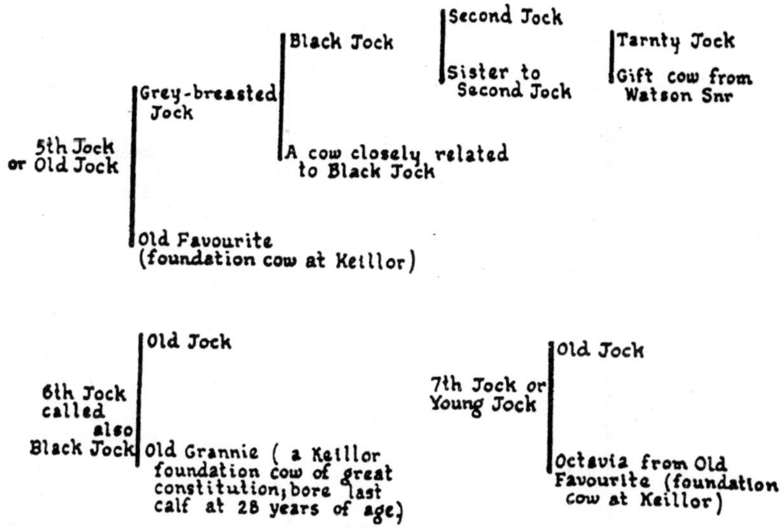

FIG. 16.
Method of developing the Keillor Jocks from Tarnty Jock to the 7th, or Young Jock.

are shown in Fig. 16. Unfortunately the available information is inadequate to do more than indicate the method.

McCOMBIE OF TILLYFOUR

The second point of interest is the method by which McCombie of Tillyfour used crosses of two relatively inbred strains to establish the very successful "Queen Mother family". McCombie of Tillyfour and William Fullerton of Ardovie were contemporary breeders. The first began in 1830 and the second in 1833. Both bought cattle from Hugh Watson, and McCombie bought a very famous cow, Black Meg, from Fullerton. The Keillor Jocks, this cow, and the bull Panmure, bred by Lord Panmure, were foundation animals for the improved black cattle.

Fullerton lost all his cattle in 1859 when pleuro-pneumonia swept the country, but McCombie was able to carry on. Before his loss, Fullerton had mated Black Meg with Captain to give a heifer, Ardovie Queen.

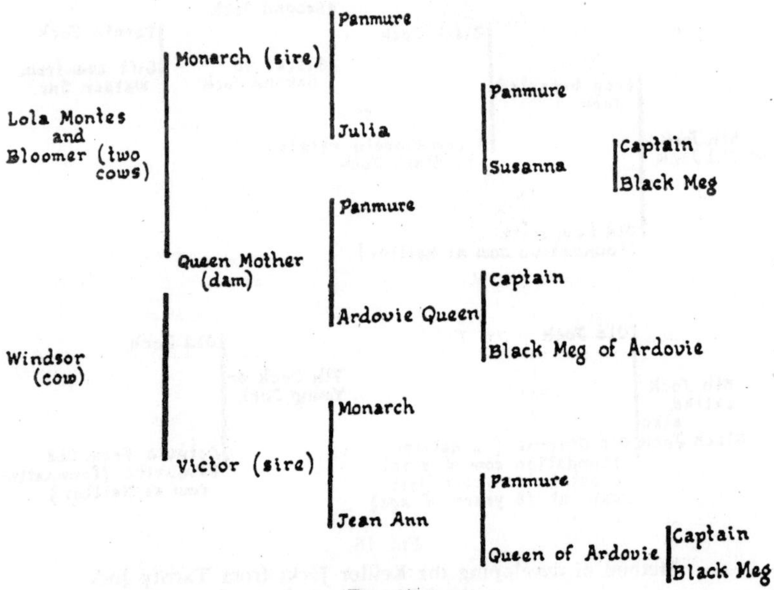

FIG. 17.

Early steps at Ardovie and Tillyfour to develop the Queen Mother Family.

She was mated with Panmure and bore Queen Mother, bought by McCombie. His method of breeding is shown in subsequent figures. It centres in his development of Queen Mother as a foundation cow. Macdonald and Sinclair (1910) quote McCombie as stating in his *Cattle and Cattle-breeders*: "Inbreeding may be pursued for a time until the type is developed; but to continue for any length of time to breed in-and-in is not only against my experience but against nature."

Fig. 17 shows the early steps taken by McCombie at

Tillyfour to develop his policy after the purchase of Queen Mother. Of the cattle shown there, Fullerton bred Queen Mother, Monarch, and Jean Ann, while McCombie bred their progeny.

McCombie then bought an inbred bull, but one not

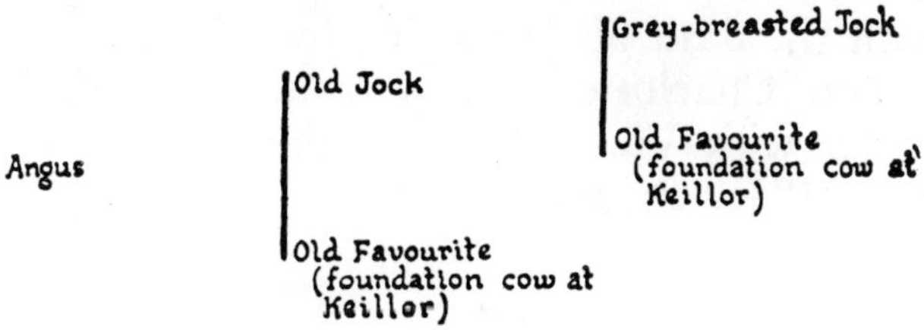

FIG. 18.
Pedigree of the bull Angus bred by Hugh Watson and used by McCombie.

related, at least closely, to these cows. That bull was Angus, bred by Watson at Keillor. He was by Old Jock and traced back to the foundation cow Old Favourite as is shown in Fig. 18. Angus was mated with Lola Montes, their offspring being the cow Charlotte. McCombie then

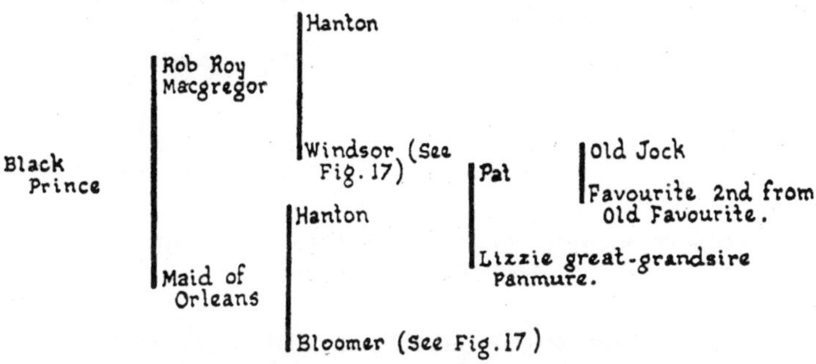

FIG. 19.
Pedigree of Black Prince as bred by McCombie.

bought another Keillor bull named Hanton. He was by a son of Old Jock from a cow that had Panmure as her great-grandsire (note just a dash of the original blood). With Hanton, Bloomer and Windsor he bred the bull Black Prince as is shown in Fig. 19. The pedigree follows on from that of the cows shown in Fig. 17.

Charlotte, the cow from Lola Montes by Angus, was mated with Hanton and the progeny, Empress of France, was mated with Rob Roy Macgregor. This is shown in Fig. 20. Note how the outcross Keillor bull Angus, with neither Panmure nor Black Meg of Ardovie blood, as far as is shown, was used, and the progeny bred back to the bull with the dash of Panmure blood.

The cow Charlotte (Angus from Lola Montes) was mated also with Victor 3rd and the progeny Crinoline was put to Black Prince (Fig. 21).

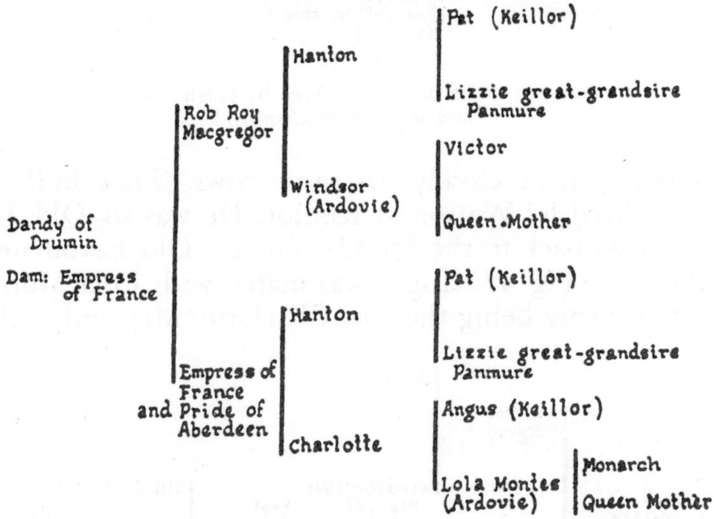

FIG. 20.
Pedigree showing McCombie approaching his synthesis of Ardovie and Keillor lines.

The available information is minimal and probably does not disclose the full extent of inbreeding that took place to more remote ancestors. However, in its present state it provides very interesting material for a study of the repeated use of animals, which, although themselves inbred, at least apparently were not closely related to those with which they were mated. The use of such animals appears to have been the kernel of McCombie's method of breeding.

Lola Montes (Fig. 17), being bred wholly from Fullerton's cattle, had a coefficient of inbreeding of approximately 22 per cent. Charlotte (Fig. 20) was the first cross of two inbred but apparently unrelated lines. From the evidence, she was not inbred. Empress of France (Fig. 20) received a greater infusion of Keillor blood from her sire Hanton, but, owing to the number of intervening individuals, her apparent coefficient of inbreeding, which was to remote ancestors, was only 11 per cent.

The same method was followed to breed Rob Roy Macgregor. The cow Windsor, his dam, was inbred to Fullerton's stock and had a coefficient of inbreeding of at least 22 per cent. Hanton, his sire, apparently was not

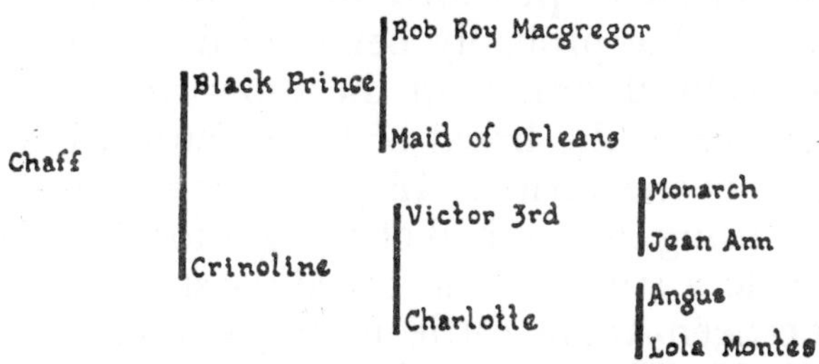

FIG. 21.
Further fusion of Ardovie and Keillor strains.

inbred. He was largely Keillor-bred, with Fullerton's Panmure as a remote ancestor appearing once. The result of this mating was a fall in the degree of inbreeding. The approximate coefficient of inbreeding of Rob Roy Macgregor was only three per cent. This was increased by mating him with his half-sister, Empress of France. It was a mating of two animals each of which combined the lines that McCombie had been cultivating, namely those evolved by Watson and Fullerton. The immediate inbreeding of Dandy of Drumin (Fig. 20) of course is due to the recurrence of Hanton. It represents 12·5 per cent of the total. The remaining 6·8 per cent in a total of 19·3 per cent is due to many lines tracing to remote recurrent ancestors.

McCombie went no further with these cattle, but bought the bull Don Fernando. He was by the bull

Windsor, bred by McCombie, and a full brother to Rob Roy Macgregor. Don Fernando's dam was from Dulcina by Cupbearer, a son of the Keillor-bred Pat from an unrelated cow. Dulcina's dam was the Keillor-bred Dora. Don Fernando was followed by an unrelated bull President 4th which had a very meritorious show record. McCombie then returned to a bull of his own breeding named Bright, by Black Prince from a relatively unrelated cow.

These methods brought McCombie considerable success in the show rings of Scotland, England, and France. The Queen Mother family, the development of which has been followed, provided many important animals, the most notable probably being Pride of Aberdeen, a cow practically unbeaten in the show ring.

Importance to the breed of these foundation animals cannot be overestimated. In 1942, Stonaker, Holbert, and Lush, using the method of Wright for Duchess Shorthorns, made a study of the breeding structure of purebred Aberdeen-Angus cattle in the United States from 1900 to 1939. They found the whole breed bore a considerable degree of relationship to animals mentioned in the foregoing. Thus they estimated Black Prince of Tillyfour to be 24·1 per cent related to the breed (practically equivalent to a grandsire's 25 per cent relation). Relationships of other important animals are Hanton, 21·3 per cent; Rob Roy Macgregor, 14·9 per cent; Grey Breasted Jock, 15·5 per cent; Earl Marshall, 14·0 per cent; Black Cap Revolution, 12 per cent. Over 60 per cent of the foundation animals were bred by Hugh Watson, Alex Bowie, William Fullerton, William McCombie, and Lord Panmure; 50 per cent in the Watson herd alone.

BALLINDALLOCH

The third and last point of interest for our purpose in this breed is the method used at Ballindalloch to maintain a high degree of excellence whilst avoiding the risks of close breeding. The herd was begun by Sir George Macpherson Grant. It is said to have contained black cattle for many years before the introduction, by purchase in 1860, of the cattle that have made it famous.

The methods used were described by C. Macpherson Grant in 1892. The breeding plan was stated to have been "to avoid inbreeding but also to avoid any absolute outcross. Breeding in line so far as possible has been adopted. Although this has not been confined to the various families, it has been applied to the herd as a whole. At the same time, individual merit has always been considered".

Ballindalloch has two outstanding families, the "Ericas" and the "Prides". Of these the Trojan Ericas, both at

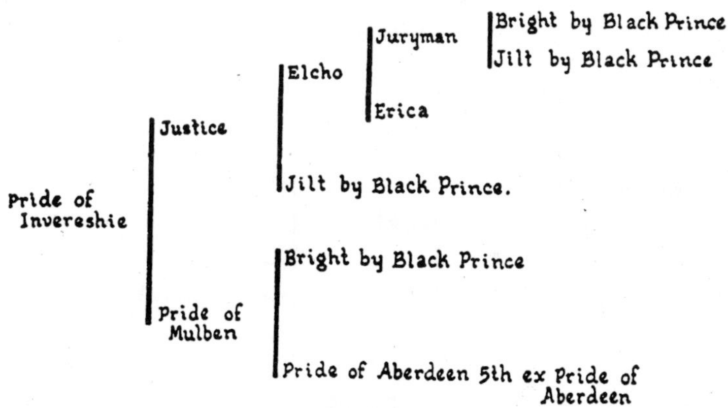

FIG. 22.
Construction at Ballindalloch.

home and in the United States, are highly esteemed. Erica, daughter of Emily by Cupbearer, Jilt, a McCombie bred cow, and Pride of Mulben, granddaughter of the outstanding Pride of Aberdeen, were the principal foundation cows. The pedigree of Pride of Invereshie (Fig. 22) will show how the concentration appears to have been to Bright and through him to Black Prince (Fig. 19). The foundation lines thus were a combination of those of Tillyfour and Keillor.

The next step is characteristic of the method used in the stud. A bull, Iliad, was introduced. He was a successful show bull by Etonian, a bull bred at Ballindalloch by Elcho from Eva. His mother was Inky, a cow largely unrelated to the herd. She was by M.P., a bull which descended on both sides from Queen Mother. It will be recalled that this method of introducing a half-outcross was used extensively in Booth's and other early Shorthorn herds. Its implications will become evident later.

Prince Inca was bred by Iliad from Pride of Invereshie. His son, Equestrian, was bred at Ballindalloch and sold; but later Bion, a son of his from Sunshine 12th, a Drumin-bred cow, was purchased and used in the herd. All three of these bulls were successful show bulls, particularly Prince Inca and Equestrian. The use made of

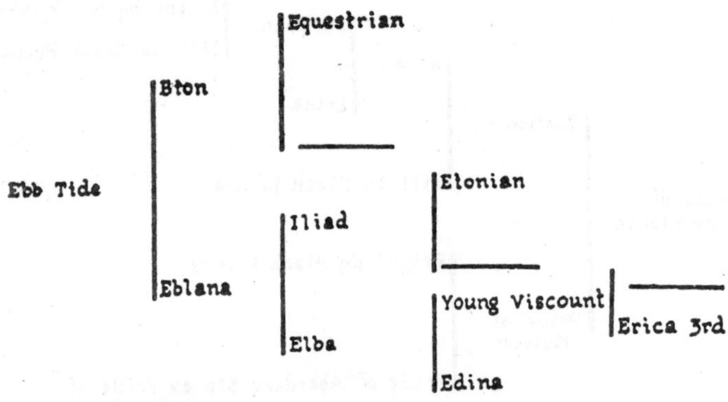

FIG. 23.
Use of Bion and half-outcrosses at Ballindalloch.

Bion is shown in Fig. 23, where he appears as the sire of Ebb Tide. In this pedigree dashes indicate the relatively new "lines of blood". It is typical of the method already quoted from the stud description published in 1892.

All the bulls were successful show bulls, and each of the three introduced a percentage of relatively new blood. Sometimes, as in the pedigree of Pride of Invereshie, more extensive use was made of bulls bred wholly within the herd. Even then inbreeding was avoided. This was possible because the herd was large and had a number of bulls. They were organized on what was stated, unfortunately without description, as the "four sires" rule. Possibly, however (see Fig. 23), had Bion, Equestrian, and Iliad been used to breed a female, instead of Ebb Tide, then she would be regarded as a proper mate for the fourth bull, Etonian or Young Viscount. There is a common plan by which a daughter, replacing Ebb Tide, would be mated back to either of the grandsires Equestrian or Iliad. This, of course, would be closer than the method suggested as possibly indicated by the "four sires rule".

In the report on the herd already quoted, twenty-seven bulls are listed as having been the principal sires between 1861 and 1892. Including animals used during the construction period of the herd, 33 per cent of these bore no immediate or close relationship to the herd cattle, 41 per cent were half Ballindalloch, half other herds, and 26 per cent were wholly of herd blood. However, most were successful show bulls and all were bred upon selected lines originating in approved cattle.

Although in the early days of the breed Watson and even McCombie practised a considerable degree of inbreeding, the history of the breed does not disclose the intense isolation of family groups that characterized the Shorthorn breed. This is well shown at an early date by Watson's use of the "new cows", Old Favourite and Old Grannie, as dams for the 5th and 6th of his Jock bulls.

In the hands of subsequent breeders the groups of cattle developed by Watson and McCombie were amalgamated relatively early. This is in contrast with the continued isolation of the Booth and particularly of the Bates strains when the herds were dispersed.

Rinderpest, pleuro-pneumonia, and foot and mouth disease, but particularly the first two of these, took heavy toll of many herds that were becoming well established about 1860. Devastations of this character may have one or other or both of two effects. The rarity of the remaining animals of particular strains might cause them to be valued unduly highly, and bred back to, without regard for possible consequences, as witness the last of the Duchess family. Alternatively, they could be used to best advantage within the surviving group. This appears to be what took place in the Aberdeen-Angus breed.

Even when Thomas Ferguson, over a period of approximately fifty years, bred relatively close, his methods, as deduced from the pedigrees of the cattle he bred, were controlled by the caution that characterizes men of the north. Other similar instances could be cited; for example, George Brown of Westertown. Macdonald and Sinclair state that

> He fixed upon a certain strain of blood and a certain type of cattle and he laboured to realize a distinct aim with such skill and method as had been displayed by the Messrs Booth and Mr Bates in breeding Shorthorns. His cattle were bred from close affinities but the dangers of excessive inbreeding were guarded against by keeping a large number of stock bulls.

It was said also of the Glamis herd that "the principle adopted was to select well-bred animals of good shapes and to mate them with specimens equally well bred with a view to improving the breed rather than by following any particular fashionably bred family". The methods used by the Macpherson Grant family were also of this nature.

CHAPTER VI

HEREFORDS: HORSES: REVIEW OF CONSTRUCTION PERIOD*

HEREFORDS

HEREFORD cattle have an ancestry little, if any, less heterogeneous than that of Shorthorns. There was also a period of anonymity for the breed, as long as that for the Holderness and Teeswater cattle. During this period Herefordshire was regarded as breeding rather than fattening country. The cattle were in the hands of farmers who used steers for draught purposes and sold them to graziers to be fattened at grass when about six years of age. This early association with specialization for a specific market exercised a similar and as strong an influence upon Hereford characteristics as the requirements of Scottish feeders had upon Cruickshank's early Shorthorns.

Red cattle similar to those now in Devon probably provided the foundation of the breed. Greys, which were common among the early herds, are considered to have been the equivalent of roans in Shorthorns. It is thought that they resulted from an admixture of the native reds and blacks with the red-eared white cattle of Wales. White faces and body markings, as seen on present-day Herefords, are attributed to infiltrations of Dutch cattle. Whole-coloured reds with brown faces, greys, reds with white faces, and cattle mottled or ticked, were all described as having been reared in the County of Hereford.

Colour is a characteristic of minor importance when compared with those economically essential for beef

* The author is considerablv indebted for information in this section to the *History of Hereford Cattle* by Macdonald and Sinclair. Vinton and Company, London.

cattle, but among Herefords it became an outstanding feature in the genetical history of the breed.

In the early days and for a period of possibly fifty or sixty years, when the movement to develop the breed was in its infancy, there were almost as many discrete strains of Herefords as there were colours available upon which to establish them. Breed historians have deplored this diversification. They regard the lack of uniformity as having restricted more general and earlier acceptance of the cattle as an established breed. Although strains were isolated by colours, the segregation was with respect to a characteristic of minor importance. The prejudice certainly was sufficiently strong to prevent cross-matings, but all the strains were being developed to comply, in every important attribute, with the same market requirement.

When the faddists became reconciled and one colour was adopted for the whole breed, its nature, fortuitously, was such that all other colours could be absorbed by it. The reconciliation occurred some time between 1849 and 1858. It provided a means whereby groups, of necessity inbred among themselves or at least isolated as discrete units having particular colours, could be outcrossed one with another. Thus faults that were becoming fixed within the groups could be corrected without the stigma of foreign "blood". In the full light of present-day knowledge no sounder plan could be advanced for the development of a breed from equally heterogeneous ancestry. It alone could provide an equal opportunity for discovering the advantages and defects of particular groups. On this evidence they could be retained or discarded. Those that withstood the trial would become relatively uniform. Matings could be made between groups to combine good points and to take full advantage of the increased range for selection.

The particular breeding methods employed during the early period of construction were comparable with those used in the same period of the other two cattle breeds. However, families of cattle do not appear to have been isolated as definitely as among Shorthorns. The Tomkins family, the Hewers, and the Jeffries were all Hereford breeders of long standing. Within their herds there was an appreciable degree of inbreeding, possibly most intense in that of the Tomkins family. However, it was the standard-coloured, large-sized Hewer cattle, and also those of the Jeffries, that eventually absorbed the mixed colours of the Tomkins groups. The Hewer herd exercised a preponderance of control because its bulls were in considerable demand for leasing. It is reported by Macdonald and Sinclair that up to thirty-five were leased to other breeders at the same time. Broadly, the breed stayed in the hands of farmers. They developed it specifically for beef production by methods similar, in most respects, to those used at Ballindalloch and Glamis for Aberdeen-Angus cattle. For this reason further extracts from the records would unnecessarily increase the material without disclosing additional principles. Those discovered can be summed up in the one word, "normality".

HORSES

Circumstances that caused development of horses in England have been indicated. Although light Arabian horses were introduced by the Normans and Crusaders, the Thoroughbred conventionally dates from the Stuart period onward. Its origin can be traced to the importation of upwards of one hundred and sixty Oriental horses that were mated with the lighter types of English mares. The imported horses were Arabians, Barbs, and Turkish, but greatest importance is attached to three of them. The Godolphin Barb, also known as the Godolphin Arabian, is stated to have been foaled in Barbary about 1724. He was taken to France, where it is said he pulled a water-cart. About 1728 he was presented to Lord Godolphin in England. The Byerly Turk was the saddle-horse of Captain Byerly. He was used in the wars of William in Ireland and was probably brought to England about 1689. The Darley Arabian was supposed to be pure Maneghi Arabian. He was bought at Aleppo, Syria, and sent, in 1706, as a present to a Mr Darley in

England. He was a dark bay about fifteen hands high and probably the most important of these three horses as a foundation sire.

The most important descendants of these Eastern horses are known as Thoroughbreds. Herod (King Herod) was foaled in 1758 and was a direct descendant of the Byerly Turk. He is said to have sired 497 winners. Cade was a son of the Godolphin Barb, and sired Matchem, who was successful on the track and as the sire of 354 winners. The Darley Arabian was sire of Flying Childers and of Bartlett's Childers. A grandson of Bartlett's Childers named Marske was the sire of Eclipse, foaled in 1764. Eclipse was one of the greatest horses in history. He won 11 King's Plates, the last (1770) as a "walk over the course", sired 334 winners, and died in 1789. Incidentally he is said to have been "thick in the wind".

There were, of course, a number of contemporary mares, many of which contributed little to the breed. Bruce Lowe considers that this honour is due to only about forty of them. However, the points of interest for our purpose are, the breed had its origin in heterogeneous stock and during the period of its construction a considerable degree of inbreeding took place. The last point is established by the pedigree of Jigg of Jiggs (Fig. 24). The pedigree was abstracted from Lady Wentworth's *Thoroughbred Racing Stock,* and can be taken as representative of the method of breeding then in vogue.

The same principle, but an alteration of method, was used by Count Orlov, a Russian in the time of Empress Catherine II, about 1772. The information is also from Lady Wentworth's *Thoroughbred Racing Stock.* She states (1938) that Count Orlov, to evolve the Orlov trotters, collected all the principal breeds at Krenovaya. Smetanka was a particularly good Barb bought from the Moors in 1775. Polkan was a son of his from a large, bony, Danish mare. He had poor action in harness. He was mated with a Dutch mare and sired Barss I in 1784. All of Orlov's trotters came from Barss I and his three sons

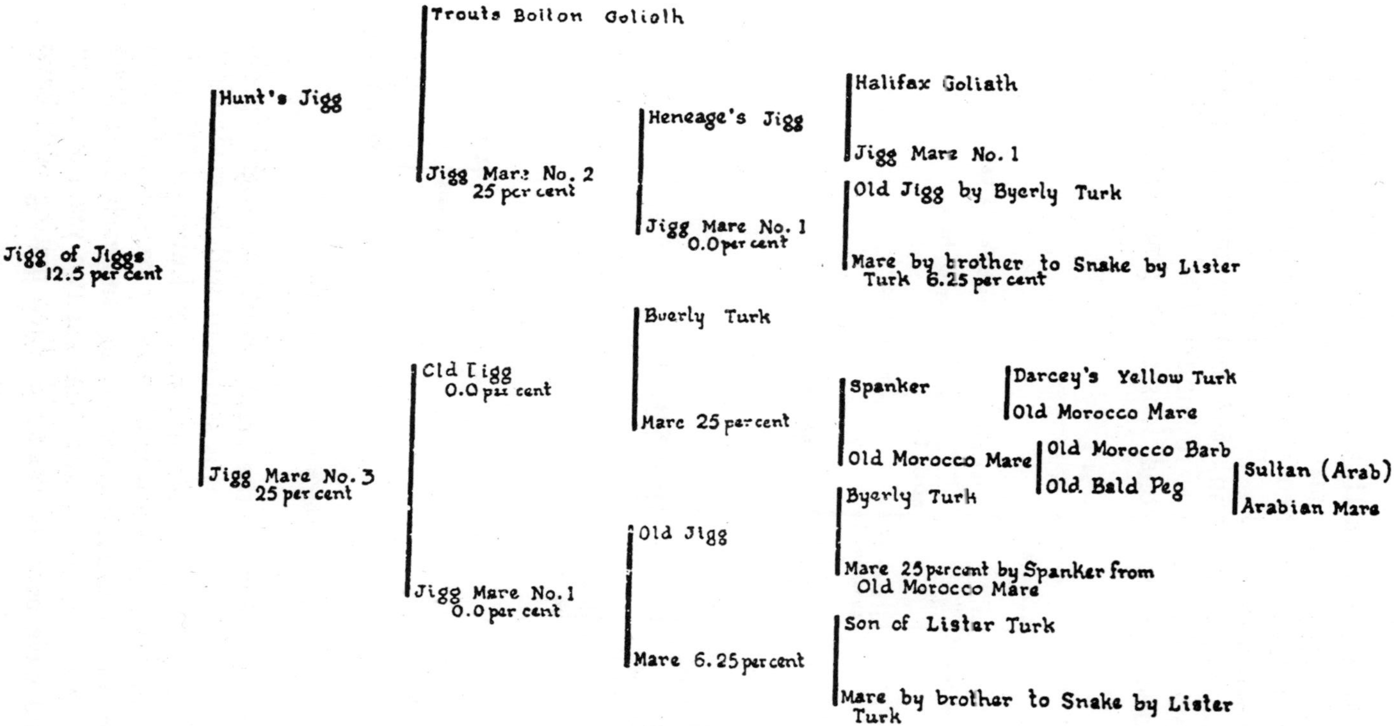

FIG. 24.

Matings in constructive period of English Thoroughbred (Jigg of Jiggs). Percentages are coefficients of inbreeding.

—Lubensoi I, Dobroi II, and Lebred I. The breed had very free action. Only mares were sold "and no one ever got good stock from them as they apparently only nicked with their own blood". The stud was purchased by the Russian Government in 1845 and the inbreeding was continued. Outcrosses were failures. Had their qualities been due to recessive conditions, as was possible from the

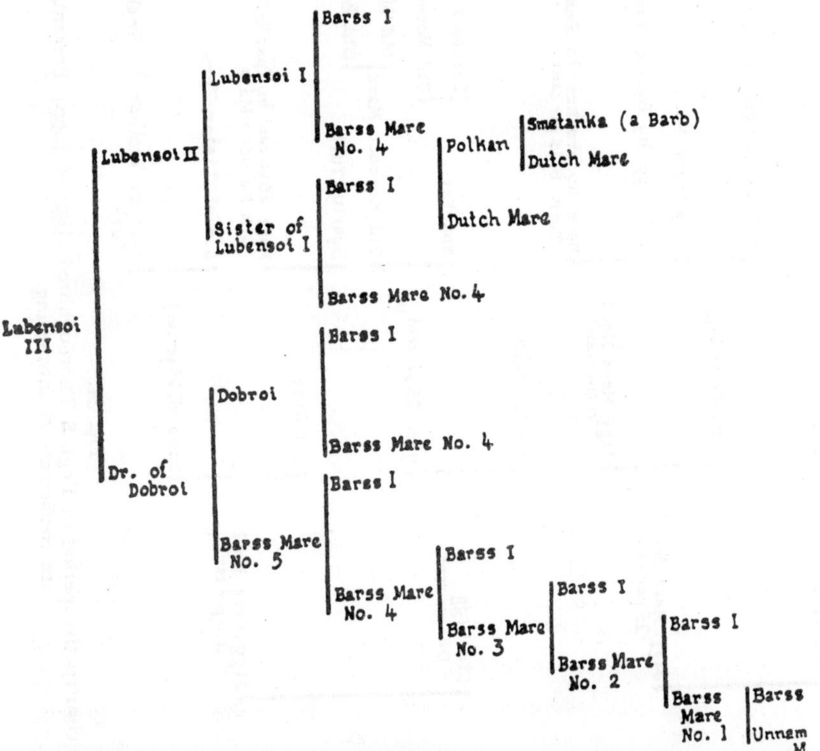

Fig. 25.

Construction period of Orlov's Trotters.

system of breeding, the result would be highly probable. The reason for this statement will be made clear later. Fig. 25 shows the pedigree of Lubensoi III as it appears in Lady Wentworth's book. If Barss I had served the first mare when two years of age and each Barss Mare from No. 1 to No. 4 when she was two years of age he would have been 14 years old when mating with Barss

Mare No. 4. There is therefore nothing impossible in this extraordinary pedigree.

Plumb (1920) states that there were heavy horses in Britain in the days of the Roman Conquest. He states also that the early history of the breed is speculative, but quotes Sir Walter Gilbey (1899) as writing that in 1654 the English Great Horses were similar to those of North German and Flemish blood. Plumb considers that there can be no doubt that the early Shire horse was of very mixed breeding. This is definitely established for those of later years. Bakewell, who was the earliest improver of Shire horses, went to Holland and returned with mares which he mated systematically with English stallions.

The origin of the Clydesdale horse is comparatively recent. He is a truly synthetic product resulting from matings between Flemish stallions and local mares and subsequently between English Shire horses and the progeny of the foregoing. Patterson, who, about 1715, imported a Flemish stallion to Lochlyoch, in Scotland, is now regarded as having established the foundation stock of the breed. The Shire was first introduced about 1860. Calder (1926) made a genetical study of Clydesdales, using the method evolved by Wright. He found that inbreeding was not great until a period dating from 1880-90. Inbreeding then and until 1912, was concentrated upon the horse Darnley, but before that there had been many scattered matings of half-cousins. After 1912 Baron's Pride was the target at which inbreeding was directed. Calder states, "There has been more inbreeding to Baron's Pride than to any other sire in the breed." Matings of daughters with grandsons and of sons with granddaughters of this horse have been common. Prince of Wales, which had for grandsire a horse Samson (grandsire also of Darnley), also has been concentrated upon. This evidence and that of Bakewell's association with Shire horses shows that the heavy draught breeds also have been "improved" by inbreeding among the progeny of heterogeneous ancestry.

REVIEW OF CONSTRUCTION PERIOD

Throughout the foregoing narrative, constant forces can be discerned implementing the various procedures, either naturally or with the knowledge and contrivance of the breeder.

Secondary environmental forces were most active. They were implicit in the requirements for cattle adapted to more ready fattening, for those better able to produce large quantities of milk, for horses of greater speed or strength, and for sheep capable of producing fine wool under conditions that were economically advantageous. They were the forces that implemented all the changes that took place. And these, in turn, were made possible largely by other environmental forces collectively classifiable as the nutriment of the animals.

Early states of all these animals provided a wide field for selection. Had they been less heterogeneous, the range would have been restricted. When they included insufficiently different types, these forms were introduced even from neighbouring countries. A mixed state in the early populations was fundamental to the procedure.

Selection within these mixed populations was possible when the selector had a preconceived objective. When his aim was earlier maturity, his observations associated that character with the peculiar make and shape or habits of certain animals. This was true also for greater speed or strength in horses and for all the other requirements. Whilst it was thus that the conception of appropriate form or "type" was evolved, it is emphasized that the conception rested upon the fact that specific economic requirements were essential. These were in turn those broadly classifiable as secondary environmental forces.

When the improvers had selected animals with the required peculiarities of form or habit, or when, their search being fruitless, they secured such types by mating and thus amalgamating different kinds, as in compensatory matings, every breeder isolated the group of livestock he had acquired. The isolation was more complete for some, but it was characteristic of all. By general agreement it was extended to all groups of similar animals, first by the rules of the Dishley Society and later by other societies of interested persons. Isolation was implicit in the conditions for registration that were the outcome of Coates's Herdbook. The movement eventually extended to all groups of animals that subscribe to similar conditions for registration and mating.

Isolation thus was general or particular. When extremely particular it applied to individuals or to relatively small groups. Then the policy could not be carried out unless related animals were mated. Inbreeding was inevitable. When less particular, but not yet general, it permitted discrete use of animals less closely related. These were used, as partial outcrosses, with greater or lesser regularity as was dictated by expediency or by the plan of operations.

Occasionally pre-existing policies of extreme isolation were controverted. This was deliberate, as with Cruickshank; or accidental, as when Hereford sub-groups were amalgamated; or cautiously planned, as by McCombie. However, in every instance where the policy of isolation was overthrown, new and wider possibilities for selection were secured among the progeny. The early state of the populations was again approximated. Nevertheless, the degree of heterogeneity in the new populations was restricted by modifications already effected and fixed, to some extent, by the nature of the earlier breeding methods. The animals differed to this extent from that of outright crossbreds; they would have provided maximal opportunities for selection but minimal fixation of most characters.

The foregoing brief review of methods used in developing certain breeds of livestock does not cover the whole field; there were many other herds and many other breeds undergoing development at the same time. It was made from the data most readily available to disclose the kind of procedure that characterized a certain period; this covers approximately 125 years, from about 1735 to

1860, or from the middle of the eighteenth century to the middle or latter half of the nineteenth.

It has been stated that biological knowledge was not extensive in the early part of that period. Nevertheless, these years marked the constructive phase of most breeds of livestock including, in many respects, our own Australian Merinos. Since then Polwarths, Corriedales, and Columbias have been developed by crossbreeding long- and short-wool sheep in Australasia and America. Certain kinds of light horses and several breeds of swine have been evolved in the United States. Long strides have also been taken toward stabilizing a Zebu x British breed crossbred, particularly in Texas, U.S.A., on the King Ranch. Broadly, however, most animal products are still secured from types developed in this era of construction.

Although biological knowledge has been progressive, it has not yet been correspondingly productive of results in animal breeding. This is incongruous, but before the possibility of closer association can be adequately discussed, it will be necessary to review briefly the growth and nature of the existing knowledge. Nevertheless, discussion can be anticipated by stating that the relative position of learning and progress in this respect is analogous to stages that mark clearing of forests to prepare arable land. Felling and burning timber are spectacular and relatively speedy. First crops can be grown among the ashes and debris, but years must elapse before the area is wholly cleared up and in full production. As a mountaineer, attacking a summit, finds the grade increasingly steep toward the apex, so, in animal breeding, the bad can be changed to good more readily than the good to very good. Moreover, either change is more simple than stabilization at the peak of improvement. Biological knowledge should be expected to facilitate the last and most difficult phases. It is thus provided with a justification and a challenge. These are to be considered.

CHAPTER VII

BIOLOGY

ANCIENT civilizations had stores of biological knowledge based upon the laborious but sound method of observation. Aristotle (384-22 B.C.) is regarded by Locy (1915) as the pioneer of such methods. In his *Biology and Its Makers*, Locy traces, from the time of Aristotle, the rise, and the fall to rise again, of the method of observation and experiment. He shows that soon after Aristotle's day men ceased observation to indulge in metaphysical speculation. Dogmatic Authority then usurped the place of knowledge. Its sway was undisputed throughout medieval times, and was not contested until the Renaissance and the Reformation changed the outlook of European peoples. Students of natural history then again returned to objective methods. How great was the reliance placed upon the voice of Authority is well illustrated by a story recounted by Locy.

In the old days dogs had been the subjects of anatomical dissections and human structures were, by analogy, assumed to be similar. Galen, who lived from A.D. 131 to 200, was still the authority on anatomy in the sixteenth century. He had stated that the human thigh-bone (femur) was curved. Vesalius (1514-64) had the temerity to question this; from his dissections he

> . . . showed that the thigh-bones are straight and not curved as in the dog. Sylvius, his old teacher, was one of his bitterest opponents; he declared that the human body had undergone changes in structure since the time of Galen and, with the object of defending the ancient anatomist, he asserted that "the straight thigh-bones, which as everyone saw, were not curved in accord-

ance with the teaching of Galen, were the result of the narrow trousers of his contemporaries, and that they must have been curved in their natural condition when uninterfered with by art."

Vesalius, in the interests of exact science, fought a battle of the offensive that was continued after his death. Locy regards developments that arose periodically throughout the centuries since then as creating epochs. He states:

It was an epoch in biological history when Vesalius (1514-1564) overthrew the authority of Galen and studied at first hand the organization of the human body.

It was an epoch when William Harvey (1578-1667), by adding experiment to observation, demonstrated the circulation of the blood and created a new physiology. The two co-ordinate branches of biology were thus early outlined.

The introduction of the microscope in the seventeenth century, mainly through the labours of Grew, Hooke, Malpighi and Leeuwenhoek, opened a new world to the investigator, and the work of these men marks an epoch in the progress of independent inquiry.

Linnaeus (1707-1778), by introducing short descriptions and uniform names for animals and plants, greatly advanced the subject of natural history.

Cuvier (1769-1832), by founding the school of comparative anatomy, so furthered knowledge of the organization of animals that he created an epoch.

Bichat (1771-1801), his great contemporary, created another by laying the foundation of our knowledge of the structure of animal tissues.

Von Baer (1792-1876), by his studies of the development of animal life, supplied what was lacking in the work of Cuvier and Bichat and originated modern embryology.

Haller (1708-1777), in the eighteenth and Johnnes Müller (1801-1858) in the nineteenth century, so added to the groundwork of Harvey that physiology was made an independent subject and was established on modern lines.

With Buffon, Erasmus Darwin, and Lamarck (1744-1829) began an epoch in evolutionary thought which had its culminating point in the work of Charles Darwin (1869-1882).

Mendel's experimental observations on inheritance, published in 1866, mark one of the most important biological discoveries of the nineteenth century, although the recognition of his work was delayed till the year 1901.

After Cuvier and Bichat came the establishing of the cell-theory (1838), which created an epoch and influenced all further progress.

Finally, through the discovery of protoplasm (1835) and the recognition that it is the seat of all vital activity, arrived the epoch (1861) which brought us to the threshold of the biology of the present day.

Step by step naturalists have been led from the obvious and superficial facts about living organisms to the deep-lying basis of all vital manifestations.

Locy's summation admirably marks the milestones. The running will be taken up about 1700 and will proceed, without regard to dates, to consider some of these epoch-making events and their implications.

ORGANIC EVOLUTION

Linnaeus, with his orderly mind, regarded similarities and slight differences among animals and plants as suitable criteria for their classification. It required Darwin's breadth of vision, however, to discern in this array and in fossil remains the origin of species.

Until then, each species was considered to have been independently created. Darwin's theory of evolution assumed continual changes with the elimination of forms not adapted to their environment. Methods by which these changes took place are fundamental to planned animal breeding because those similar, but possibly of lesser degree, are also responsible for variations that give rise to breeds and strains of animals.

Organic evolution postulates a single cell as the ultimate form of life. At the beginning, this cell underwent relatively minor changes and gave rise to other but slightly different forms. Some of them were better adapted to survive than either the original cell or its other modifications.

In forms capable of modification, this went on throughout millions upon millions of years during which the earth's crust was formed and reformed. Living organisms of today are the types that survived; others are represented principally by fossil remains. For every advantageous modification that took place there were vast numbers that were not of that character. Many forms were incapable of modification and perished; thus whole species were eliminated in the struggle for existence.

Amphibia were the first animals to live on dry land. Like their modern representative the frog, they spent their early life in water, breathing by means of gills; but they made use of lungs during adult life. Reptiles arose from one of their unspecialized branches. The earliest known reptiles resemble the primitive Amphibia so closely that it is difficult to discern their differences. However, the Reptilia eventually reached a much higher grade of structure and almost completely superseded the Amphibia on dry land.

Although the reptilian type once was successful and widespread, it has failed in the struggle for existence; but it gave rise to other types that replaced it. Birds are descended from some primitive reptile. They underwent all kinds of adaptations, and most lines survive to the present day. Mammals, the class to which man and all his domestic animals belong, arose earlier than the birds but also from reptilian stock.

Quite independently of birds, and on different lines of specialization, the Mammalia have acquired a four-chambered heart and a self-regulating mechanism for maintaining their normal body temperature independent of that of the surrounding environment. They soon diverged from primitive egg-laying types, which are now almost extinct, but still survive in monotremes like the platypus. Adopting the advantageous method of nourishing their young in the mother's womb during early life, placental mammals spread rapidly over the earth. They replaced the lower reptilian forms and diverged in many adaptive directions. They have, however, suffered severely in the struggle. Large groups have vanished altogether, while others are on the verge of extinction. The marsupials, once widely distributed, remain only in Australia and as scattered forms in America.

Hoofed herbivorous mammals started very early, in Eocene times, from primitive forms no larger than a fox. At that time they had five-toed feet, which were reduced during processes of adaptation. They are divided into two great groups, one with an odd and another with an even number of toes. Horses and asses belong to the former group; sheep, cattle, and pigs to the latter. These comprise most of the domestic animals.

RECAPITULATION AND GERMINAL CONTINUITY

The theory of Organic Evolution has received an enormous amount of supporting evidence, not least important of which is that given by the law of recapitulation. It states that every animal, during its development from a single germinal cell, climbs its own genealogical tree. Of course the stages are not complete but, extraordinary as it may appear, the gill-breathing state of fishes is extremely well defined in embryos of birds and mammals. There is implicit in this statement another, known as continuity of the germ-plasm. By it such phases of past lives are attributed to something in the germ-cell which it carries to succeeding generations. Before this can be discussed it will be necessary to consider the structure and reproduction of a single cell. It is the form of life from which all others have developed, and it is the unit of construction for every living tissue.

STRUCTURE AND REPRODUCTION OF THE CELL

A typical cell has a wall that encloses a substance, called protoplasm, in which there is a central body or nucleus. The nucleus has an outer membrane and consists principally of threads of material which readily stain when exposed to colouring agents during the preparation of specimens for microscopic examination. For this reason it is called chromatin material.

When a typical cell divides to reproduce another, the chromatin goes through a well-established series of movements. In the course of these it breaks up into component parts known as chromosomes, of which the number is constant throughout a particular species. Equal numbers of the halved chromosomes move, or are pulled, to either end of the cell, which constricts between the two new nuclei. This gives rise to two new cells. It is important to note that the material of every chromosome is halved and that the number of chromosomes in the nuclei of the new body-cells is the same as that in the mother cell. For example, Ahmed (1939) has shown that sheep have 54 chromosomes; thus, every time a body-cell divides in a sheep, each new body-cell also has 54 chromosomes. This is so, regardless of the breed of the sheep.

Division of germinal cells is different. It is so devised that each new cell contains only half the number of chromosomes found in body-cells of the same animals. For example, for sheep, where there are 54 chromosomes in the body-cell there are only 27 in the germinal cells of either sex. The process is called reduction division and the double state (54 for sheep) is referred to as diploid, while the reduced state (27 for sheep) is the haploid state. In the male, each remaining group of 27 chromosomes provides a nucleus for another of the immense number of spermatozoa, but in the female the residue is expelled from the cell by the production of polar bodies. Extrusion of particular chromosomes is controlled wholly by chance and is an essential part of the mechanism of inheritance.

When reproduction takes place in domestic animals, each parent provides one germinal cell with a complement of chromosomes in the haploid or half-sized state (27 in sheep). When the cells, one from each sex, fuse, the zygote, as the new cell is called, has the double or diploid state, that is, 54 for sheep, made up of two lots of 27. A new individual, progeny equally of the two parents, develops from the fertilized egg cell or zygote.

Very early in the growth of this new individual, specialized tissue begins to form the testes or ovaries of the embryo so that at birth, and derived from the germinal tissue of its parents, the offspring has its own organs capable, in their turn, of producing the germinal tissue appropriate to its sex. In this way continuity of the germ-plasm is ensured from generation to generation and with it the chromatin material which research has disclosed to bear the units of inheritance. These units are called genes and the movements described, by which chromosomes carry genes from cell to cell and from generation to generation, constitute the chromosome mechanism.

CHROMOSOME MECHANISM AND ORIGIN OF SPECIES

When this method of inheritance was worked out and widely accepted, it became necessary to reconcile it with the idea of evolution. The conception provided an adequate explanation for the law of recapitulation, but it was completely at variance with the possibility of progeny inheriting characters acquired by parents during their lifetimes. There are obvious instances where this does not happen; for example, lambs born from sheep with amputated tails do not themselves have short tails. Nor is the position affected by the fact that sheep's tails have been shortened by this method for very many generations. In this connexion it is interesting to note that lambs have been bred without tails or with tails less than one and a half inches long. This has been done by cross-breeding with the introduction of new genes into standard populations. However, there are observations with regard to acquired characters which at first sight suggest that they can be inherited. For instance, when animals became better adapted for survival, it was necessary that improved parents should pass the advantage to their progeny, otherwise these, being in no better state of fitness than the original type, would die.

These and other facts have been explained by showing that chromosomes, which are relatively few, act as vehicles for immense numbers of something to which the name "gene" has been given. Fruit flies (Drosophila) of various kinds have been subjected to very close and extensive genetical examination. They have only four chromosomes but it is estimated that at least 6000 pairs of genes are carried upon them. If anything like a similar proportion were probable, the number in sheep would be in the order of 80,000 pairs. Genes have not been isolated and examined chemically or morphologically. In certain forms, however, their size has been estimated. They are said to be somewhat larger than virus units and very much smaller than any of the well-known micro-organisms. They should not be regarded as independent pieces of a jig-saw puzzle which produce their effects by fitting into appropriate places. They appear to be chemical substances which have shapes similar to those well recognized in organic chemistry for the structure of molecules. An appreciation of their chemical nature will simplify many outstanding questions. Genes are always at fixed positions on the chromosomes and are carried from cell to cell upon their own particular portions of chromatin material. The movements of this material have been well established as taking place normally by a series of figures as regular as those of the old-time dances. Transport of genes by chromatin material and the labile state of the gene itself are most important in the chain of causes that make the theory of evolution reconcilable with inheritance by the chromosome mechanism.

Although normally animals reproduce progeny like the parents, upon occasion different kinds arise. These differences may be within the range expected as normal variations or they may be something quite new. In the latter case, the change is occasionally abrupt, but more often it is gradual, increasing through several generations until the particular character becomes completely different.

This spontaneity is attributed to changes in the gene structure of the parents. The change may result from the taking up of abnormal positions by threads of chromatin material that are carrying large numbers of genes, or they may be due to alteration in the structure of particular genes.

The aberrant movement of threads of chromatin material would give rise to great differences between parent and offspring, while modifications of individual genes would probably cause minor changes. When these are repeated, however, the changes become cumulative and more marked. This is probably the most common form of variation of this nature.

The end result of such nuclear changes is that, whereas for many generations specific effects could be attributed to the action of particular genes, all at once these cease to "run true to form". Instead of producing horned cattle for instance, they produce polled beasts. Or, as actually happened, chickens from normally-feathered fowls were hatched naked. In lower forms of life, which have a nuclear structure and behaviour similar to that of higher forms, shells of snails, for example, might suddenly grow with spirals from left to right instead of right to left as they did formerly. Alternatively there might be a series of minor changes which would eventually produce the maximal end result.

These changes, which originate in the gene complex, that is, in the aggregate of genes and their vehicles of chromatin material, are called mutations, and the modified animal or character that arises from them is spoken of as a mutant type or a mutation. Mutant types breed true for the alterations. Further, it has been shown that, by the use of certain chemicals and of X-rays, mutations can be produced at will.

Mutations thus provide an explanation of how differences originated. There was an ancestral type which for a long time bred true to its parental form; but suddenly it gave rise to progeny that were different. Alternatively, the difference may have been the end-product of a number of minor changes, the result of a gradual process. In either case, the difference would have to be advantageous for the mutant type to persist. Perhaps it made the animal more fertile or more resistant to cold or heat, so that it reproduced more freely or survived a change of climate that killed other kinds. Variation arising from all possible combinations of gametes and controlled by chance will be discussed later as a further possible cause of differential survival.

The foregoing discussion is abridged and considerably simplified. For example, changes due to aberrant movements of chromosomes have not been elaborated. Further, at many points detail of the mechanism is still being examined (Dobzhansky, 1941). However, although it is well established and satisfies many theoretical requirements, the mutation hypothesis does not appear wholly satisfactory for the conditions of practical animal husbandry.

The conception is most satisfying and acceptable when it refers to changes that have taken place throughout geological time. Sceptics, however, remember the story of the earthling who was being introduced by Saint Peter to Heavenly ideas. In an endeavour to impress local time and values upon the new arrival, Saint Peter told him that a penny was worth £100,000,000 of earthly money and a minute of heavenly time equal to 7,000,000 years on earth. When the earthling tried to borrow a penny, Saint Peter asked him to wait a minute!

The critical point of the mutation conception, in its application to constructive animal breeding within the life-span of the ordinary man, is the occurrence-rate of mutations. Sceptics have watched plants, and particularly insects, breeding freely and extensively. They know that the mutation rate in closely observed populations of Drosophila (fruit-fly) with some 6000 pairs of genes is not greater than about 0·7 per cent of the germinal cells produced by an individual. They know that generally the mutations gave rise to characteristics less advantageous than those of the orginal wild types; that most frequently they produced minor effects and so belonged to cumulative types of mutation which are slow to produce marked changes.

This information is common to many animal geneticists who have joined the sceptics in asking for something more. They express their queries as statements. They say that unless the mutation rate in large animal groups is fairly rapid, it would have small chance of explaining, for example, the present Australian Merino in contrast to the original kind. Breeders of improved beef cattle have a similar comment to make with regard to early fleshing ability of their animals and to their earlier maturity.

These changes, and others that will readily occur to the reader as having taken place within his own lifetime, require a more realistic explanation than that presented by the mutation hypothesis. They can be interpreted more satisfactorily as the shuffling and regrouping of existing genes or the introduction of new genes by crossbreeding—two causes that will be discussed in later sections.

Some authorities suggest that although acquired characters are not inherited as such, environment may play a part in modifying the germ-plasm. This may be confused with the reaction of the organism as a whole. All individuals are the result of interaction between two sets of factors; those of inheritance and those of the environment. The normal structure of an organism is that developed under the usual complex of environmental stimuli. Any divergence from the normal structure may be due either to some change in an otherwise constant environment or to an altered environment acting on the unchanged genetical constitution. The real difference between characters that always appear and those which do not, is that the former depend on stimuli that are always present while the latter depend upon stimuli that may sometimes be absent.

Goodrich (1912) illustrates this conception admirably by stating:

> The inheritance might be compared to a musical instrument, the stimuli to the player and the organism and its characters to the music produced. What particular tune is produced will depend on the player, but the range of possible sound will be limited by the structure of the instrument.

It can also be illustrated by work of Fourie (1939) in South Africa. He investigated a condition which produced photo-sensitization, skin trouble and eventually early death in Shorthorn and Freisian cattle in that country. From test matings of affected and related stock he showed the condition to be inherited as a recessive. It was thus part of the heredity of the animals. But it had not been observed among their parental stocks in either Great Britain or Holland. The condition required South Africa's bright sunlight, i.e. the African environment, to become demonstrable.

The hypothesis of modification of the germ-plasm should not be summarily dismissed. However, it does not affect materially the truth and usefulness of the theory that variation, crossbreeding, isolation, and selection are fundamental to the differentiation of new types.

SCIENTIFIC METHOD

While much of the foregoing is based upon observation, some is speculative. This is characteristic of scientific procedure. Direct observation provides material from which an hypothesis is advanced. This properly should be tested by planned experimentation. Without this, the hypothesis is an expression of opinion having no greater weight than that of its perpetrator's suitability to express such an opinion. When proven by the method of experimentation the hypothesis can be accepted as a theory. When a theory has stood the test of time and been verified from every angle by experimental and other trials, then it can be regarded as truth or as a natural law.

Although it was founded upon widespread observations it was obviously impossible to prove, by direct experimentation, Darwin's hypothesis for the origin of species; the lifetime of man was, and is, the limiting factor. However, a great volume of supporting evidence, some of which has been discussed, has established the hypothesis. The same method can be traced in establishing the chromosome mechanism. The series of movements has been observed upon innumerable occasions, but the hypothesis that chromosomes act as vehicles for units of inheritance relies upon supporting evidence for verification. The method is implicit in the nature of the material. It is characteristic of many biological problems, but not so frequently of those in chemistry, physics, or other natural sciences.

The procedure is evident also in the foundation material of the new science of genetics. Although now an extension of many older established sciences, it had its beginning in the work of Gregor Mendel, an Austrian monk who became interested in the inheritance of characters in garden plants. Unfortunately his work, although published in 1865, received no attention until the phenomena he described were rediscovered by three scientists working independently some thirty-five years later.

CHAPTER VIII

MENDELISM

MENDEL's work was not that of one accustomed to microscopic study, but that of a keen observer of macroscopic objects and a logical thinker. Mather (1942) emphasizes his faculty for logical thought by stating:

> Gregor Mendel formulated two laws of heredity from which our theory of particulate inheritance has developed. But to do so he devised an experimental technique for investigating genetical differences and this is, perhaps, his major achievement. Without the technique he could not have made his own discoveries. With the technique he could solve his problems and, what is more, later geneticists could test, modify and extend his laws into the modern theory. Mendel's technique ensured the development of genetical science.

Mendel's technique depended on the use of single differences. His classic experiment was with two kinds of peas, one dwarfed, the other tall. When he cross-pollinated these two he found that all their progeny were tall. When these tall crossbred plants were self-fertilized, their progeny were of the two original kinds in the ratio of three tall to one dwarf. These second generation plants, when self-fertilized, showed by their offspring that the dwarf type bred true for shortness and that some of the tall plants bred true for height. These were one-third of all tall second-generation plants. The other two-thirds of the tall plants of that generation had progeny with a 3:1 ratio of tall to dwarf.

When he understood the individual behaviour of this character, and of smooth and wrinkled seeds, Mendel carried his work further, into combinations of the two. He eventually added others and "drove", as it were, all these characters in from "single harness to four-in-hand". His technique for doing so was unique, and deserves the commendation bestowed upon it by Mather.

Mendel learned that each of the characters with which he was working, when considered independently, behaved as tallness and shortness had done. But when they were regarded collectively, different ratios were discovered. When two characters were considered, the second generation plant that always bred true for both parental characters that had the same nature as "shortness", that is, a recessive condition, occurred as one of 16 (15:1). A similar plant in the second generation with three parental characteristics was one of 64, a number which is four times as great as 16, which in turn is four times as great as 4, the number which included one such plant (3:1) in the simplest case.

Mendel rightly emphasized the particulate nature of these inheritances. This is most important. It is completely at variance with the conception of blending inheritance, and is a most valuable contribution to an understanding of heredity as disclosed in domestic animals. These are not blends, but composites of more or less independently inherited characters. That these are ill-defined forty years after the rediscovery of Mendel's work does not invalidate the conception. Emphasis, therefore, is placed upon the independent inheritance of specific characters, although under linkage and other conditions some may not behave in this manner.

Linkage and the peculiar kinds of inheritance for which it is responsible were unknown to Mendel. He was very fortunate in his choice of the material with which he worked because, for the greater part, the characters disclosed complete dominance and their controls acted simply, without confusing interactions. Because of this he included dominance in his findings, but later work has disclosed that dominance of one character over another is not an essential feature of heredity. In addition to his technique Mendel discovered (i) that genes segregated, or separated, again and were unaltered in their effects by their close association with other genes while in crossbred or hybrid forms, and (ii) that their combinations and recombinations were at random, that is, controlled by chance. He distinguished clearly between the factor or gene itself and its effect in the plant, thus establishing the conception that the appearance of an animal or plant is not necessarily an indication of the kind of progeny it will reproduce.

From these beginnings the science of genetics has developed. In discussing its extent and nature Lush (1937) stated:

> Linkage, the chromosome basis of heredity, the conception of dominance "as only an incident in the chemistry of growth by which one gene produces a much greater effect than its allel (mate) upon the developing individual", and other extensions of Mendel's work, are now, by common consent, included in the definition of Mendelian inheritance.
>
> While the inheritance of any character remains unknown it is not possible to state definitely that all inheritance is Mendelian in this broad sense. However, many cases of inheritance which at one time were considered to be other than Mendelian have since been shown to owe their differences to the very large number of genes involved or to complicated interactions between them. Broadly, scientists now agree with the statement of East and Jones in 1916, namely: "Mendelian heredity has proved to be the heredity of sexual reproduction; the heredity of sexual reproduction is Mendelian." However, it is generally conceded that most characters depend upon a large number of genes.

POST-MENDELIAN PERIOD

During the forty-odd years that have elapsed since Mendel's work was rediscovered and fully appreciated, an enormous quantity of material has been accumulated. Characters dependent upon a few genes were speedily resolved in many species of plants and animals but, for one reason or another, problems of characters depending upon large numbers of genes, that is, polygenic characters, have been deferred. Nevertheless, it is abundantly evident that information with regard to these is a major requirement for further progress, because most characters of economic value in animals are obviously of this nature.

Genetical research has been influenced by the relative suitability, as material, of quick-breeding insects. This has led to much greater fundamental knowledge of the mechanism of inheritance, and has also increased disputation as to the causes of species origins, but it has not attacked the most important problems of animal breeders.

For our present purpose, therefore, contributions made by the science of genetics since Mendel can be regarded as (i) a number of first principles derived from laboratory investigations designed to learn "more and more of less and less", (ii) fairly complete knowledge of the inheritance of a number of simply inherited characters, and (iii) results from a series of desultory "flank attacks" which have replaced "frontal attacks" upon the problem of polygenic characterization.

Geneticists since Mendel should not be condemned offhand for taking these lines of least resistance. They were early faced with the knowledge that, as the number of desirable characters increased, and as the number of genes upon which they depended became larger, so also was there an increase in the size of the population necessary to secure a "pure line". This will be shown in the following paragraphs, but it is emphasized that it is this knowledge, the low rate of reproduction, the high cost of domestic animals, and, prior to 1932, an inadequate technique, that have been the principal restricting factors. Even plant geneticists, who begin with self-fertilized cereals, are restricted to mass action by the complexity of polygenic characters. Thus there still remains an inconceivably enormous quantity of work to be done before a mechanism of inheritance can be established with certainty for every individual character of animals or even for those that are most desirable. Until this happens, animal breeding must retain much that is artistic and empirical rather than scientific and exact. Nevertheless, within the last ten years, Fisher, Wright, Haldane and others have interested themselves in statistical procedure. This has been evolved to investigate the small effects of each of many gene groups or causes when non-heritable fluctuations occur. It has also made possible appraisal of probable effects by procedures based upon biological first principles. With these, problems of multiple-factor (polygenic) inheritance at least can now be engaged in "patrol activities".

PURE LINES

Attention has been drawn to the fact that the population, that is, the number of individuals, necessary to disclose particular Mendelian ratios increased directly with the number of genes involved. However, it must be realized that the ratios will only be disclosed when the numbers involved are sufficiently great. For example, a character like polledness in cattle that is customarily regarded as depending upon the action of one pair of genes, that is, as a unit-factor character, can be expected to demonstrate a ratio of three polled to one horned when first-cross (progeny of polled x horned) animals are bred together. However, if there are only four progeny, the ratio may not be apparent. There would be a greater chance of finding it among 44 animals, and a still greater chance if the total were say 400 or more.

The point is raised to disclose the very great size that second generations must reach before there can be selection adequate to isolate animals homozygous for several characters that depend upon a large number of independent genes. The animals that will breed pure for parental types after crossbreeding are in the ratio of 1:2:1 for a unit-factor character. Each of the "ones" represents an animal "pure" for either of the parental characters. Table 1 shows the rate of increase and states the general formula where n is the number of pairs of independent genes or factors involved. It also shows the number of animals which, because they have different genetic constitutions (genotypes), are capable of leaving dissimilar progeny. It will be seen that the minimal population size is four raised to a power (multiplied by itself as many times) represented by the number of factor-pairs involved. Genes or factors are carried on the chromosomes. If each pair of chromosomes carried only one of the genes involved, the population size necessary to find one animal that would breed true (be homozygous) for any particular combination of 30 characters would be 4 to the 30th power, or more than one billion billion animals. The number of animals capable of leaving unlike progeny would be greater than 206,000 billions. The size of these figures is indicated by the fact that there are only about 600 million sheep and approximately the same number of cattle in the world.

TABLE I

NUMBER OF GENE PAIRS AND SIZES OF POPULATIONS

Number of pairs of genes	*Size of population*	*Number of different genotypes*
One pair	$4^1 = 4$	$3^1 = 3$
Two pairs	$4^2 = 16$	$3^2 = 9$
Three pairs	$4^3 = 64$	$3^3 = 27$
Four pairs	$4^4 = 256$	$3^4 = 81$
n pairs	4^n	3^n

Such figures as those in the table are based upon the simplest possibilities. They would be influenced and considerably reduced by linkage, by chromosome misbehaviour, and by gene interactions. Further, all breeds of domestic animals within the same species must have a very large number of fixed genes common to all. For example, all cows have similar anatomy and physiology although they may differ in many other respects. It is the genes responsible for these differences that should be thought of as suggesting a value for *n*. Nevertheless it is obvious that to attempt to secure "pure" domestic animals by continued inbreeding for a combination of many characters, except under very exceptional circumstances, is impracticable; hence also concentration in programmes of planned inbreeding upon one character at a time.

The possibility discussed has reference to animals "pure" in the genetical sense and not to those "pure" in the everyday sense of "pure breeds" like Herefords, Shorthorns or others; to an exact state in the combinations of discrete independently acting genes. It is necessary to emphasize this point. It establishes the futility of wishful thinking based upon incomplete knowledge with regard to the possibility of securing animals genetically pure for all of a large number of characters. However, the position as a whole is not equally hopeless. The heredity of all animals results from the action of aggregates of genes; and the management of these can in many respects be compared with the management of sheep. Except for a few stud rams, no sheep-owner works his sheep one by one. He treats the flock as a whole, drafting and classing them in lines. That is the real lesson of the discussion of "pure lines". Genes, as units of inheritance, for the greater part must be considered as aggregates with a few "tops" that can be handled separately. Within an aggregate of genes, which is spoken of as the genotype of the beast, there are many groups that vary in size. Broadly, each of these groups, some with only one pair of genes and others with up to possibly scores or hundreds, produces specific effects. These are the independently inherited characters evident in the animal. With our present state of knowledge we cannot draft out many of the gene groups, but we can work out means by which they can be controlled, to some extent, as aggregates. The methods are comparable with those that handle sheep in large flocks but not individually. The contributions of geneticists since Mendel can thus be considered as (i) the principles of inheritance and knowledge of the action of few discrete genes or gene groups the equivalent of the stud rams, (ii) an appreciation of gene aggregates to be regarded as the flock sheep and (iii) the idea of characterization as a complex of forces which can be analysed (drafted into) to disclose its component factors and their relations.

CONTROL BY CHANCE

Probably the most valuable conception of all was contributed by Mendel. It is control by chance or, as it is more generally stated, random assortment of genes. This begins in the halving of the number of chromosomes to give rise to germinal cells. Either of the paired genes on the chromosomes may be pulled into one or other half of the nucleus. In the formation of the female cell the rejected half no longer has any possibility of usefulness. It is thrown out of the cell. In the male cell each half provides the haploid number of chromosomes for one or other of two sperm cells. These enter into competition with millions of others, but only one will customarily enter each egg-cell. Some conception of the size of these

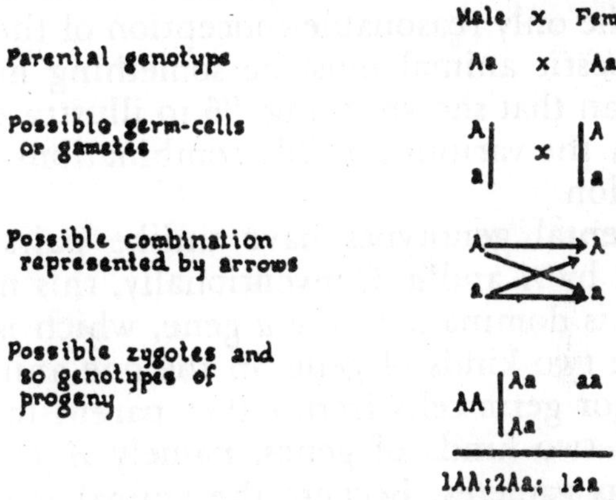

FIG. 26.
Results of control by chance at fertilization.

cells can be secured from the statement of an American geneticist that the sperm-cells in man responsible for all persons on earth would fit into two ordinary pin's heads while the human egg-cells requisite for the same population would fit into a quart bottle.

All genes are paired. This is so even when they have complex interactions. At fertilization each paired gene has every possibility of uniting in every combination. This is shown by the arrows in Fig. 26, in which conventional letters are used to represent genes. For simplicity, only one pair is shown, but it should be realized that few characters, and certainly no animal, have anything approaching an equally simple genotype. For example, hornedness and polledness in cattle are repeatedly referred to as expressions of one pair of genes. In a general way this is true also for sheep but, in that species, the position is complicated by sex associations. However, if the reader will cast back over the kinds of horns he can remember, they will become evident as having been shelly, tight, wide, strong, weak, downturned, snaily, scurs, butts, etc. In this case, although the presence or absence of horns is broadly dependent upon the action of one pair of genes, obviously there must be a host of others that are associated to produce such differences. Similarly, the only reasonable conception of the genotype of any domestic animal must be something much more complex than that shown in Fig. 26 to illustrate diagrammatically how the various possible combinations take place at fertilization.

The parental genotypes have unlike pairs of genes represented by *A* and *a*. Conventionally, this means that the *A* gene is dominant to the *a* gene, which is recessive to it. These two kinds of gene, in the one animal, make it possible for germ-cells from either parent to have one or other of two kinds of genes, namely *A* or *a*. These are shown as gametes. Because the animal is capable of giving different kinds of eggs, that is, germ-cells or gametes, it is spoken of as being heterozygous, a term derived from the Greek words *heteros* meaning different and *zygos* an egg.

In sexual reproduction, male and female cells fuse to fertilize the ovum or female cell. There are literally millions of spermatozoa available, any one of which might carry an *A* or an *a* gene. Further, the female cell may have cast out an *A* or an *a*. Chance alone controls which of the possible combinations does take place, but all are shown by the arrows. When any gene meets its mate they achieve the paired state in the zygotes shown.

Mendel's experiment with tall and dwarf peas was similar to the scheme discussed, although he started with the two pure breeds, tall (*AA*) and dwarf (*aa*). These gave an *Aa* zygote. This first-cross plant was tall. He found, however, that although it looked tall it did not breed uniformly that way, as is shown in Fig. 26. When two characters act in this way—the one completely overshadowing the other in a cross of the two—the one that disappears is called recessive to the other which overshadows it. The one that remains evident is referred to as the dominant. Dominance of this nature is not demonstrated nearly as often as Mendel supposed, but much importance will be attached later to recessive conditions.

Broadly, if animals beget progeny which in particular respects are unlike the parents, it is an odds-on chance that the characters then disclosed are recessive conditions. The two kinds, dominants and recessives, require different procedures for selective breeding either for or against. They comprise two gene groups which, like weaners and wethers in a flock, require different treatment.

SEX-LINKAGE

The idea of linkage, and, particularly in the poultry industry, of sex-linkage, is another Mendelian unit that can be handled separately. The incidence of particular sexes is spoken of as the sex ratio, that is, the number of males to number of females. Many attempts have been made to control this ratio without success. However, among the chromosomes, one pair is known to be associated with sex. In domestic animals other than fowls the pair can be identified in one of two forms. These are known as *XY* or *XX*. The *XX* chromosomes are always present in females and the *XY* in males. This is the double or diploid state as in the zygote, but when the chromosomes become halved, as in the gamete, either parent can provide an *X* but only the male can supply a *Y*. Such chromosomes are shown in Fig. 27 as moving from the parents to their progeny. When, by chance, as is shown in Fig. 27 (*a*) the *Y* and the *X* form up in the zygote the animal that develops from that fertilized egg will be a male—in the figure, a colt. If, however, and equally controlled by chance, a gamete carrying an *X* chromosome from the sire fuses with one carrying an *X* chromosome from the dam, they will form up in the zygote as *XX*. It will give rise to a female—in the figure (*b*), a filly. The shaded portions represent all the remaining chromosomes of the nuclei of both parents and offspring.

Genes reside on and are carried by all the chromosomes. It is obvious, therefore, that if any particular gene were on the *X* chromosome, its effect when in the recessive state, would always be evident in males although usually masked by the dominant of the pair in females. That is what sex-linkage means. Recessive genes carried on the *X* chromosome are linked with sex because a male can receive an *X* chromosome only from its female parent.

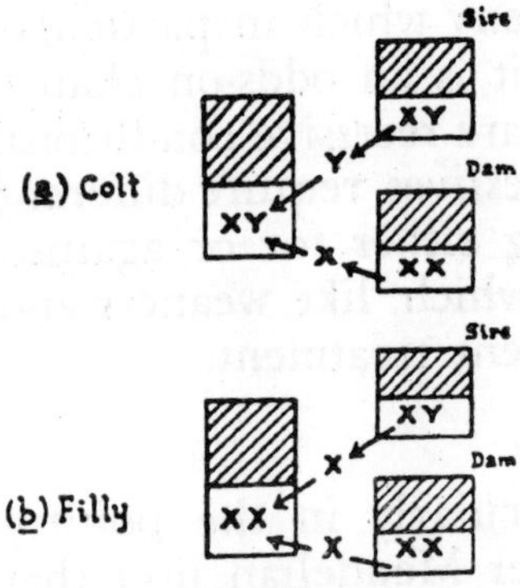

FIG. 27.
Sex-linkage.

In poultry breeding, female chicks are more desirable than males, and sex-determination among day-old chicks has become a common practice. Incidentally, the method involves opening the cloaca, which in birds is the common passage for both urine and faeces, and recognizing, in it, a rudimentary male organ. In some breeds of fowls, male chickens are recognizable by their characteristic feathering, the genes responsible for the colouring being carried on one of the sex chromosomes which in their species are referred to as *ZW*, this being the female combination. Several kinds of distinctive colouring and sex have been discovered and are in use. A typical combination is barred and black. When a black cock is crossed with barred hens, as in

Plymouth Rocks, all the male chicks are barred and all the hens come black (Hammond, 1940).

Buchanan Smith (1937) has put forward an hypothesis that sex-linkage in dairy cows has an influence on milk-yield but not on percentage butterfat. The matter is controversial, because milk-yield is polygenic and it is unlikely that all the genes involved would be on the sex chromosome. However, a very large volume of data has been examined most painstakingly and, as Buchanan Smith points out: "The original observation of the possible existence of a sex-linked factor was made more than nine years ago, and in the interval no direct evidence has been led against this observation despite the fact that several workers are engaged on the problem." He also states:

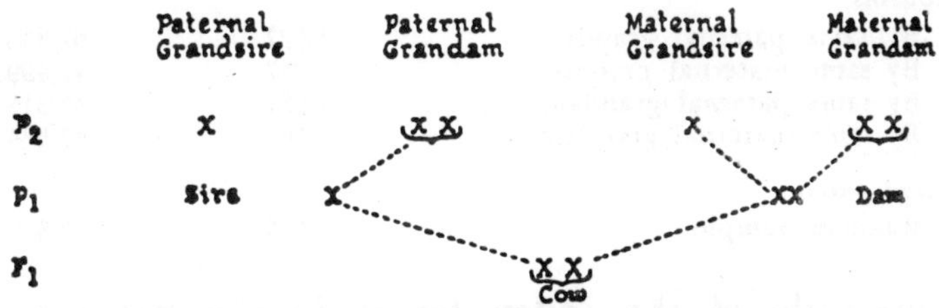

Fig. 28.
Showing diminished influence of paternal grandsire.

> The practical implications . . . are that, in the selection of a bull for a herd, undue emphasis should not be placed on his sire. It is to the good that he should be a proven sire . . . but as regards the desired sex-linkage gene, this can only be obtained by the bull from his dam, and therefore the yield of the dam of a bull is of prime importance. . . . The other implication refers to inbreeding. If sex-linkage is important, then for inbreeding to be effective, it cannot work by the concentration of the blood of a famous sire through his sons.

The transmission of sex-linked characters over three generations is shown by Buchanan Smith, Fig. 28, and the data in Table II were presented by him in support of his hypothesis.

TABLE II

BUCHANAN SMITH'S DATA

SHOWING DIFFERENCE IN MILK-YIELD OF PAIRS OF DAIRY SHORTHORN COWS FROM THREE HERDS ACCORDING TO THEIR RELATIONSHIP CLASSIFICATION

Relationship groups	*Number of individuals*	*Indices*
1. Full sisters	191	38,644
Half-sisters		
2. By the same sire with dams paternal half-sisters	2,506	32,822
3. By the same sire with dams unrelated	11,903	44,155
4. By the same dam with sires unrelated	215	59,577
Cousins		
5. By sires which are full brothers ..	248	33,329
Half-cousins		
6. By same paternal grandsire	3,627	62,856
7. By same maternal grandsire	8,737	47,499
8. By same paternal grandam	1,021	45,318
9. By same maternal grandam	667	47,684
Unrelated pairs		
10. Random sample	7,573	78,357

Magnitude of the index figure in the last column is indicative of the difference. It is least in classes 1, 2, and 5, greatest in unrelated pairs 10, and nearly as great in 6, as would be expected from Fig. 28. These data are also of general interest in that they show that blood-relationship is associated with similarity. This point is fundamental to the use of inbreeding and will be emphasized later. Already it has been referred to in the isolation of breeds and herds.

It is well known that Bruce Lowe places extraordinary emphasis upon female descent in thoroughbreds, and more recently another hypothesis has been put forward by Keylock. This also emphasizes the female descent, but more particularly importance of the grandam. If there is any true difference, that is, one completely dissociated from uterine or post-natal nutriment, in the effect of male or female parent upon the heredity of their progeny, it can only be associated with sex-linkage as shown or with gene interactions. In all other additive respects contribution of the male or female parent is potentially equal. These two hypotheses are opinions, as defined. They are subject to the same general criticism as sex-linkage in milk-yield, but they have nothing like the weight of evidence to support them.

INDEPENDENT BEHAVIOUR OF GENES

Independent assortment of genes is fundamental to all crossbreeding, including compensatory matings within pure-bred livestock. It is the outstanding cause of variations within strains which in turn give rise to new types. The phenomenon will be illustrated by almost its simplest form—that of crossing two head of cattle which have, among many others, two independently inherited characters each known to show complete dominance.

	Male		Female
Parental Type	RRHH	x	rrhh
Gametes	RH		rh
Zygote	RrHh		

FIG. 29.
Parental character in polled and horned cattle.

The chosen animals will be a black polled Aberdeen-Angus bull and a horned Hereford cow. In these, black and red, polled and horned, white face and coloured face, are three paired conditions, each of which is inherited independently and each of which depends upon the action of one pair of genes. Such cattle have many other groups of genes, all of which will be discarded, including those for one of the characters mentioned. This is done for simplicity.

The parental characters being considered, therefore, are the first two pairs; black and red, polled and horned. These are written conventionally, in the diploid or paired state as *RR, rr* and *HH, hh*. These are the paired genes before reduction takes place. The male is black (*RR*) and polled (*HH*); the female is red (*rr*) and horned (*hh*). These two types are homozygous, that is, they can only give rise to gametes of one kind, each gamete containing one gene for each condition, that is, half the pair.

When the two gametes fuse to form the zygote which will give rise to their progeny in the F_1, or first filial, generation, the genes will be in a paired state. They will be as shown above (*RrHh*) and, because black is dominant to red and polled to horned, the beast will look black and polled. It is incidental, and has nothing to do with the two characters under consideration, that the first-cross also has a "baldy face". However, the fact that this does occur without interfering in any way with what is happening to body-colour and to horns is an example of the particulate nature of inheritance. The importance of this point cannot be overestimated. Along with the white face are a host of other characters, like milk-yield and temperament, all of which are discarded for the moment.

	Male		Female
F1	RrHh	x	RrHh
Gametes	RH		RH
	Rh	x	Rh
	rH		rH
	rh		rh

FIG. 30.
Possible kinds of gametes in progeny of black polled and red horned cattle.

It is assumed that more than one first-cross beast is bred and that some of these are mated together. All have black bodies and are polled, and all have arisen from zygotes represented by the formula *Rr Hh*. We have seen that there are all possible ways in which these paired genes may break up during reduction division to form ova and spermatozoa, so we must represent as gametes all possible combinations of each gene. The *R* and the *H* are one, the *R* and *h* are another, the *r* and the *H* are a third and the *r* and the *h* are the fourth. There are thus four kinds of gametes possible in both parents. When there are sufficient numbers, so that there are many ova and many millions of spermatozoa any one of which may

	Female Gametes RH	Rh	rH	rh
RH	RRHH	RRHh	RrHH	RrHh
Rh	RRHh	RRhh	RrHh	Rrhh
rH	RrHH	RrHh	rrHH	rrHh
rh	RrHh	Rrhh	rrHh	rrhh

FIG. 31.
Possible combinations in the second generation.

fuse with any ovum, there will again be all possible combinations. This was shown earlier by arrows, but the simplest way to work these combinations out is with a checker-board. The four kinds of male cells are written down one side of a large rectangle and the four female cells along another. Then each subdivisional rectangle is filled in with the appropriate combined formula. The *RH* on the top of the vertical or male side can meet and fuse with any one of the ova represented on the horizontal or female side. The two gametic formulae are placed in the appropriate rectangle, being written again in the paired state because each is a possible zygote that will give rise to an individual in the second filial generation. It is interesting to remember that genes for white face and all the other gene groups are behaving similarly. The series of zygotes under consideration will be seen to be 16 in number, i.e. 4^n where $n=2$, or 4^2 (=16).

When these are classified, there is only one *RRHH*, the homozygous parental male type, and only one *rrhh*, the homozygous female type used in the first mating. In addition to these there are others in which the genes have combined differently. Because black (represented by *R*) is dominant to red (represented by *r*) and because polled (represented by *H*) is dominant to horned (represented by *h*), it is possible to predict what the calves from these zygotes will look like. This is shown in the accompanying table (III).

It will be seen that only four kinds are homozygous. If it were desired to continue breeding black polled cattle or black horned cattle, or red polled cattle or red horned cattle, it would be advantageous to learn which of the beasts apparently conforming to those descriptions would actually breed true for their state when mated together.

The red horned lot is easy, but not because there is only one. If there were say 20 × 16 cattle or 320 head involved, there would be 20 red horned beasts. These two conditions, red and horned, are recessive, and thus are demonstrated only when in a homozygous state. The only way to find which of the nine black polled cattle

TABLE III

TABULATION OF APPEARANCE AND GENETIC CONSTITUTIONS OF SECOND GENERATION OF BLACK POLLED X RED HORNED

Number	*Description of beast*	*Genetic Constitution*	*Number*	*Number homozygous*
9	Black polled	*RRHH*	1	1
		RRHh	2	—
		RrHH	2	—
		RrHh	4	—
3	Black horned	*RRhh*	1	1
		Rrhh	2	—
3	Red polled	*rrHH*	1	1
		rrHh	2	—
1	Red horned	*rrhh*	1	1

is truly of that type (*RRHH*) is to make a trial mating. In animals, such trials are generally made by mating back to the recessive, which in this case is red and horned, *rrhh*. If all nine types that look like one another, that is, are black and polled, are mated with red horned, all except the *RRHH* type will leave *some* progeny which *show the recessive condition*. The second one, *RRHh*, will give some horned calves, the third one, *RrHH*, will give some red calves, and the fourth one, *RrHh*, will give some red horned calves. This method of testing for a recessive condition in a sire that is intended for extensive use is of fundamental importance.

IMPLICATIONS OF INDEPENDENT BEHAVIOUR

Now let us look at these results. When the two parental types, black polled and red horned, were mated, the first-cross looked like one of the parental types. If the parental types, however, had been black horned and red polled, the first-cross would have been black and polled, and so unlike either parent. Further, among the F_2 generation (Table III) there are two new types, some of which would breed true for their appearance. Recombinations of this nature give rise to the variants that make selection of types possible after crossbreeding. It was gene movement of this kind, but in a much more complicated state, which provided the early breeders with their range of selection. It also made it possible for Colling to make compensatory matings and for Bakewell to introduce Flemish mares to English Shire stallions and so get something different from either parental type.

Instead of these simple formulae a frightfully complicated one could have been written; for example, there are kinds of wheat in which the red colour varies from very light red to deep red, depending upon the number of genes for colour that are present. If some of these kinds were crossed, the independent movement of any number of the genes would result in correspondingly small or large variations from the parental types. That probably is what happened in the origin of species. It will be remembered that, among other causes, normal variation was mentioned but withheld from discussion.

It is evident also from a study of the appearance of the animal, which would result from some of the zygotes, that this is no criterion of how it will breed. For example, most of the black polled cattle in Table III would sometimes breed other than black polls. The appearance of the animal is one thing, called its phenotype; but its genetic constitution, called its genotype, is another and quite different thing. Sometimes, when the animal is homozygous—e.g. (Table III) *RRHH*, *RRhh*; *rrHH* and *rrhh*—and mated with others of its own kind, it will breed as expected from its appearance or phenotype. On many more occasions, however—in this simple example, in 12 out of 16 instances—this will not take place. This explains why so many sires that are not regarded highly on their appearance breed excellently, particularly rams with their many characters. It is also the explanation for emphasis upon progeny-testing as the best method of determining the breeding worth of a sire.

It is interesting to observe also that a recessive gene can remain hidden, or latent, indefinitely—until it meets its appropriate mate. For example, the *RrHH* type could be mated with *RRHH*, *RRHh*, and *RRhh* types indefinitely without leaving a red calf, and as a result would be regarded as pure for black. However, if at any time it or some of its progeny met a mate of the other types shown in the table, some red calves would be left, provided the numbers were adequate.

Elsewhere (Kelley 1942) the pedigree of a litter of Border Collie pups has been used to illustrate this point. Red is recessive to black in that breed. The pedigree shows that a black and white bitch mated with a black and white dog bore two red and white pups in a litter of five. For three generations on the sire's side and for four on the dam's side all the ancestors had been black and white. The recessive genes remained hidden or latent through these generations but became demonstrable when they were appropriately paired. For such progeny, the expression "throw-back" is frequently used.

CHAPTER IX

POLYGENIC OR MULTIPLE-FACTOR CHARACTERS

IMPORTANT as the foregoing points are for logical thought upon animal breeding, they cannot compare in this respect with the conception that most characters of domestic animals, and particularly those of economic worth, result from the action and interaction of very large groups of genes. Among these, dominance is frequently ill-defined or absent, and for the greater part they seem to act cumulatively. This apparently gives rise to a blending inheritance quite different from that in which characters readily segregate in the manner previously discussed. Multiple-genes of this nature are the flock sheep of our earlier simile. Realistic discussion of planned animal breeding resolves itself into consideration of how such genes behave in the mass and how they can be controlled in that state, with little specific knowledge of the nature of their actions.

If the reader reflects upon the stature of man he will realize that, although there are some dwarfs and a few giants, the vast majority are somewhere about "average height". This is true for most characters, whether they be of the mind or of the body, of men or of animals.

A Frenchman named Binet originated a series of questions by which people could be classified in respect of the association between intelligence and age. The classification was expressed as an Intelligence Quotient or I.Q. Woodworth (1934) states, with regard to these tests and to the distribution of intelligence in the population:

> Individual differences are as pronounced here as in any dimension of human activity. About 60 per cent of children have an I.Q. between 90 and 110, differing little from the exact average, while 20 per cent are below 90 and 20 per cent above 110.

Woodworth then gives the distribution in detail as in the following table:

TABLE IV

I.Q. RATINGS FROM WOODWORTH'S DATA

I.Q. rating	–70	70 79	80 89	90 99	100 109	110 119	120 129	+129
Number per cent of examinees	1	5	14	30	30	14	5	1

Moving to an altogether different series of data, there are those of Ritzman (1923). He worked on the inheritance of size and conformation in certain sheep, Southdowns and Rambouillets (Merino). Measurements were made of parts of the animals. These were expressed, for his purposes, as ratios, for example, height divided by length, and so on. He used many of these ratios, but the one to be abstracted from his data is that for loin width over trunk length. It is shown in Table V for both parental types and for their first and second cross (F_1 and F_2) progeny. Attention should be concentrated upon the distribution of the measurements, particularly in the parental and first-cross types. The nature of the distribution should be compared with that from Woodworth's data.

Another, and the last of these series of data, is from botanical studies. It is presented in the form of two frequency tables, chosen because the numbers are sufficiently large to demonstrate the distribution. It is emphasized that in all cases the greater the number of observations, the more complete will the distribution be.

These distributions show that, with regard to all of the characters, the populations examined were classifiable into groups that bunched up toward the middle but thinned out at the extremes of the range of characterization. This is true for any specific character of this nature. By this method any group of similar animals, for example,

TABLE V

DISTRIBUTION OF RATIO LOIN WIDTH TO CHEST LENGTH IN SOUTHDOWN AND RAMBOUILLET SHEEP AND THEIR CROSSES

(Data from Ritzman)

Figures show number of individuals within the range

Generation		*Range of Ratios*																*Number of individuals*
		17	*18*	*19*	*20*	*21*	*22*	*23*	*24*	*25*	*26*	*27*	*28*	*29*	*30*	*31*	*32*	
P	South-down						2	3	2	4	7	9	12	5	4	1	1	50
P	Ram-bouillet	2	5	8	10	9	9	5	2									50
F_1	First cross			2	6	3	10	10	4	1								36
F_2	Second gener-ation					2	5	4	8	8	6	3	2	3				41

TABLE VI

FREQUENCY OF DIFFERENT LENGTHS OF BEANS MEASURED BY DE VRIES

Length in millimetres	8	9	10	11	12	13	14	15	16
Number of beans	1	2	23	108	167	106	33	7	1

TABLE VII

FREQUENCY OF DIFFERENT TYPES OF BEECH LEAVES DETERMINED BY K. PEARSON

Number of veins in leaf	10	11	12	13	14	15	16	17	18	19	20	21	22
Number of leaves of each type	1	7	34	110	318	479	595	516	307	181	36	15	1

all cattle or all of one breed, can be classified as being low scorers, high scorers, or of the great majority. These descriptions can be converted for purposes of selection into the bad and very bad, the good and very good, and the vast majority or general average. If the unit of classification were a composite of all the characters of an animal, for example, shape, size, staple length, density of fleece, and wool character of a sheep, the classification and the distribution would probably be the same. However, the preceding discussion has emphasized repeatedly why each character should be isolated and regarded independently. Having in mind existing methods of mass classification with the whole of the animal as the criterion, no solitary statement of greater importance could be made.

THE NORMAL CURVE OF DISTRIBUTION

Data in any of the preceding tables could have been presented graphically. The first step is to arrange them in class intervals, as has been done with Woodworth's data. This makes a series of frequency polygons (Fig. 32, top). If the class intervals had been extremely small and points joined, the shape of the graph then would have been a regular and continuous curve (Fig. 32, bottom). This curve is called the "distribution curve". Theoretically, it extends to infinity at either end of the range.

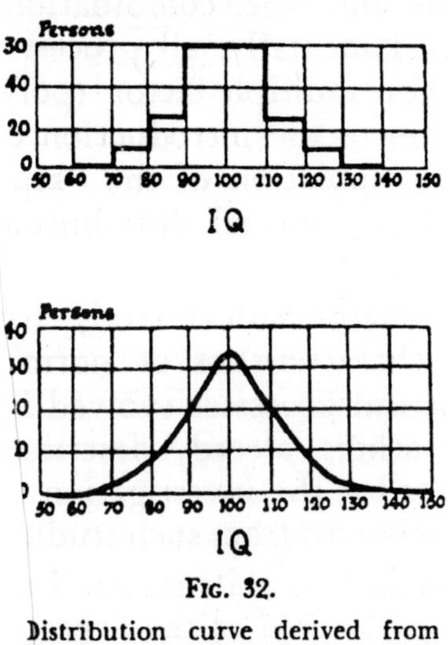

FIG. 32.

Distribution curve derived from Woodworth's data, Table 4. *Top*, frequency polygons; *bottom*, distribution curve. Frequency is expressed as per cent per step of 10 points I.Q.

Curves can be prepared from all such data. They may vary in shape, but when the observations have what is referred to as a normal distribution, the curve which represents them graphically is spoken of as the normal curve, or the curve of normal distribution. In this kind the area below the curve represents the whole population and is proportional to it. Almost any characters, such as fleece-weight or milk-yield, that fall within the province of animal breeding will provide normal distributions, provided only that the population being examined is unselected and sufficiently large. Sometimes, however, the apex leans to one or other side. The curves are then spoken of as being "skewed". They occur when inheritance of the character is influenced by dominance or by gene interactions. Normally distributed observations indicate the simply additive types of inheritance that are being considered.

Emphasis placed on the important part played by chance in controlling mechanisms of inheritance will have prepared readers for the statement that all occurrences in any game of chance, if graphed in the foregoing manner, will also provide normal distributions. This can be verified if several coins are tossed many times and the occurrences of all tails, all heads, and other combinations are recorded and graphed. Fundamentally, all processes of selective animal breeding for multiple-factor (polygenic) characters are closely akin to the introduction of coins with two heads or to manipulation of the "kip" which carries the coins so that the normal distribution can be upset.

A considerable amount of mathematical study has been concentrated upon the characteristics of normal distributions. Indeed, the statistical processes evolved by Fisher, Wright, Haldane and others, already described as providing greater possibilities for the investigation of polygenic characters, have been evolved from such studies.

The first property of the distribution is its mean. This is commonly known and frequently used as the average, being determined by dividing the sum of the values of the observations by their number. Most observations within a series have values different from that of the mean. The differences in these values are referred to as deviations from the mean. They may be positive or negative, depending upon whether the value of any observation is greater or less than that of the mean. The deviation is zero when there is no difference.

By mathematical processes beyond the range of our discussion, a statistic has been determined, called the standard deviation and represented by the Greek letter sigma (σ). This statistic is important because it defines the limits of the great majority or general average of a population that is normally distributed. Fig. 33 shows a normal curve with the base on both sides of the mean marked at points each distant one standard deviation from the mean. They are shown in the figure as -1σ and $+1\sigma$.

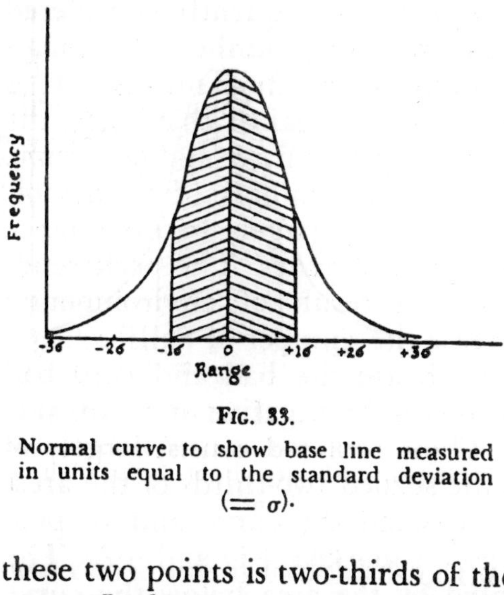

FIG. 33.
Normal curve to show base line measured in units equal to the standard deviation ($= \sigma$).

The area below the curve and bounded by the perpendiculars to it from the base and erected at these two points is two-thirds of the total area below the curve. It thus represents that fraction of the total population supplying the data. This area, which is shaded in the figure, represents the general average or great majority.

There are also two unshaded areas below the curve. Each of these represents one-sixth of the total area. This fraction (⅙ × 2) is the proportion of a normally distributed population that is outside the range of the general average. The plus and minus signs represent members of the population better or worse than the general average.

NORMAL DISTRIBUTION AND CULLING OF UNSELECTED POPULATIONS

It is emphasized that the conception of this shaded area, and the sixths lying outside and above or below it, as representing the composition of an unclassed flock or herd, is fundamental to the proper appreciation of polygenic characterization and the methods by which it can be controlled when the genes are additive. The one-sixth should be compared with the 20 per cent stated by Woodworth (p. 91) to be below ninety points.

It suggests that minimal cullings would be one-sixth, or 16 per cent of the whole flock or herd. The remaining 14 per cent of the 30 per cent so frequently considered to be necessary depends upon the number of animals available and the most desirable residual number. It is influenced also by selection for or against characters, the inheritance of which is not simply additive and polygenic. They may depend upon small numbers of factors or, although polygenic, be influenced largely by dominance or by pronounced interaction of genes. The occurrence of undesirable animals as a result of environmental causes also tends to increase the number of culls.

Broadly, the culled sixth are the bad and very bad members of the population on the negative or minus side of the general average. Those retained consist largely of fair average sorts from the shaded two-thirds of the area, and of the good lying between the curve and perpendiculars erected at the points marked $+1\sigma$ and $+2\sigma$. The very good are represented by the area below the curve and above the points on the range marked $+2\sigma$ and $+3\sigma$.

Hammond (1940) quoted percentages of butterfat determined by Wriedt among first-cross and back-cross cows derived from Jerseys and Red Danes. The data are reproduced in Fig. 34. If those of the first-cross cattle are examined it will be seen that the range of percentage butterfat among 108 head was from 3·4 to 5·5 per cent. The mean is 4·39 per cent and the standard deviation is plus or minus 0·43 per cent. If this figure were added to and subtracted from the mean, two positions would

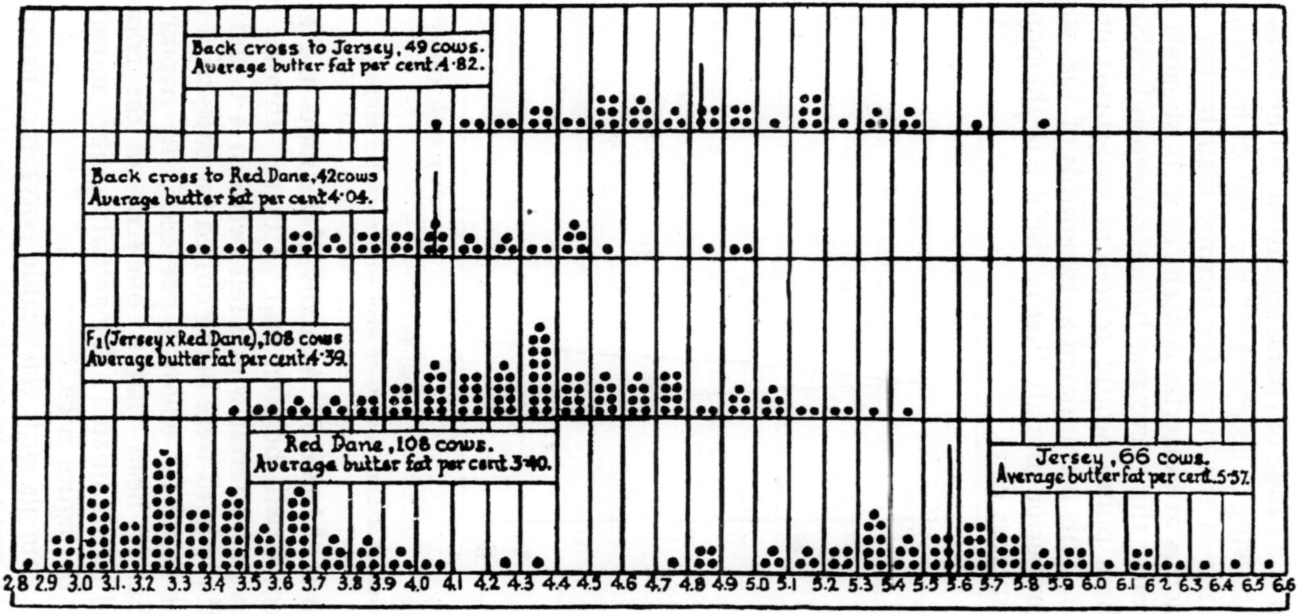

FIG. 34.

Percentages of butterfat in Jerseys, Red Danes and their crosses (data after Wriedt).

be fixed in the range of characterization. They would be at points showing 3·96 and 4·82 per cent of butterfat.

Association of the curve for normal distribution with such characters as butterfat content of milk suggests that the area below the curve and between perpendiculars to it from these points would represent two-thirds of the population. This is translated as two-thirds of 108, which was the number of animals, that is, the size of the population of first-cross cows.

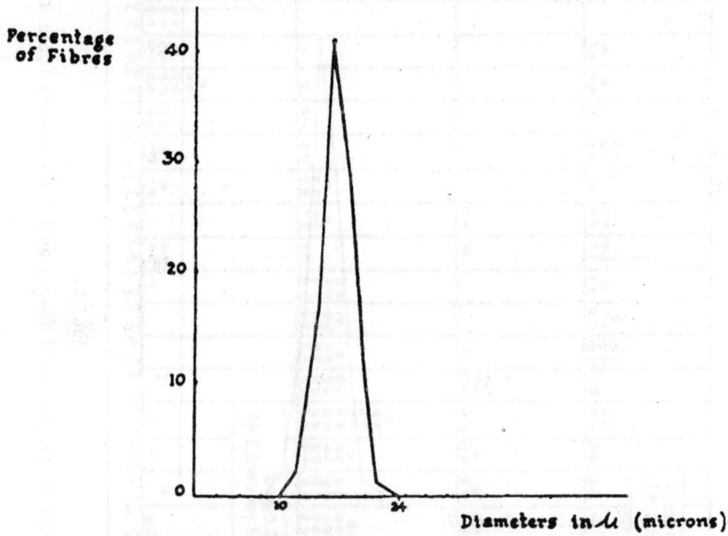

FIG. 35.

Distribution of fibre fineness of wool with restricted range.

Two-thirds of 108 is 72. Actually, if the spots in Fig. 34 are counted, it will be seen that 71 of the animals had percentages of butterfat which fell between 4·0 and 4·8 per cent. If the herd were being culled for cows with percentages of butterfat below 4 per cent, the hypothesis suggests that 18 head, that is, one-sixth of the population, would be eliminated. The actual number shown by the data in the figure is 19.

Broadly, animals with classifiable polygenic characteristics that place them in the lower sixth of the curve for unselected stock provide the culls. The object of planned breeding is to restrict the area under that part of the curve; in other words, to restrict the number of animals providing data to place them in that position when classification takes place. The ideal curve for the occurrence of any character would be one on an extremely restricted base. However, such a degree of uniformity might, in some instances, be undesirable for other reasons. Nevertheless, it does occur, as witness the accompanying curve (Fig. 35) for distribution of diameters of wool fibres in a particularly uniform sample.

REGRESSION TO THE MEAN

Wriedt's data (Fig. 34) also disclose another very important point. It is the swing, or movement, of the averages of the five groups of cattle. This is shown dia-

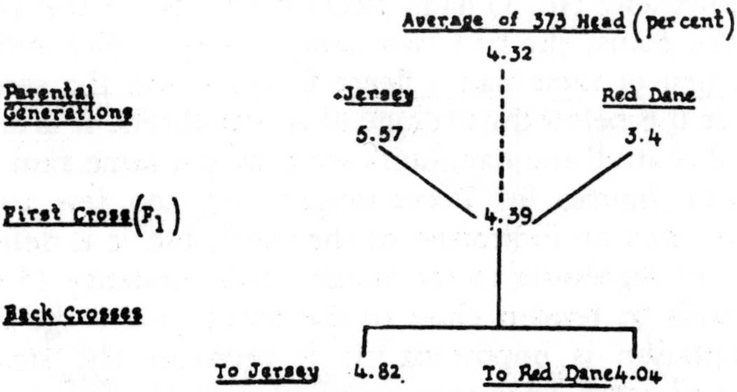

FIG. 36.
Movement of average per cent of butterfat in Wriedt's data.

grammatically in Fig. 36. The average percentage of butterfat for all five groups comprising 373 head was 4·32 per cent. The average percentage of the first cross was 4·39. The back-cross individuals swung, one higher and one lower, towards the mean of the recurrent parental types, but they did not reach as high or as low a point as the average percentage of either pure breed. There was a pull towards the general average of all such cattle that limited extent of the swing in either direction.

The same tendency can be seen in the following figures

(Table VIII) for fleece-weights. The data were arranged to isolate weights of fleeces from a group of ewes all of which had daughters shorn at the same time. Eight ewes satisfied the condition, All were the same age and all had suckled a lamb during the year when they grew the fleeces. None of the daughters had borne a lamb, but all had grown their fleeces in the same year, and in the same year as their dams. All the daughters were by the same sire. There were 11 of them; 9 were 2 years old and 2 were 3 years old. All the sheep were depastured on the same small property.

The average fleece-weight (greasy wool) of all the dams was 8·3 lb., that of the daughters 7·8 lb. Departures from these means by the dams and by their daughters are shown in Table VIII. The figures should be read as follows: ewe No. 59 had a fleece 0·3 lb. below the average of the dams; she had two daughters Nos. 580 and 299; the first of these had a fleece 0·55 lb. and the second a fleece 0·8 below the average of all daughters. It is emphasized that all the daughters were by the same ram.

The figures for fleece-weights are too few to give other than an indication of the trend, but it is definitely that of regression to the mean. This tendency to swing towards or remain close to the mean or average of the population is important. It is probably the strongest natural force operating upon characterization that depends upon large numbers of genes, that is, upon polygenic characters. It is critical in mass selection based upon phenotypes and, unless specifically attacked, definitely restricts progress by that method.

Surveys of production by daughters of dairy bulls compared with the production levels of the dams show how strongly regression to breed average controls the position. Ward in New Zealand has published the following figures (Table VIIIA) for the percentages of bulls which improved, maintained, or lowered production in herds of cows with different levels of production.

TABLE VIII

FLEECE-WEIGHTS OF DAMS AND DAUGHTERS BY A COMMON SIRE, SHOWING DEPARTURES FROM MEANS OF FLEECE-WEIGHTS FOR THE TWO GROUPS

Average fleece-weights	*Dams*	*Daughters*	
	8·3 lb.	*7·8 lb.*	
Identification numbers of the dams	*Departures from the mean of all dams (lb.)*	*Identification numbers of the daughters*	*Departures from the mean of all daughters (lb.)*
59	—0·3	580, 299	—0·55, —0·8
51	—0·3	306, 491	+0·2, +0·7
12	—0·3	444, 291	+0·7, —0·3
45	—0·55	410	+1·7
95	+1·2	415	+0·45
63	—0·8	396	—0·8
15	+0·45	588	—1·05
4	+0·7	316	—0·05

Taking 350 lb. of fat to be average production of the dams, it will be seen how few of the bulls were advantageous mates for cows with higher levels. Conversely, it will be seen also how few bulls were disadvantageous when mated with cows having below-average levels of production. Only 6 per cent of the bulls were advantageous on the highest level compared with 73 per cent on

TABLE VIIIA

Average Production of Cows in Herd (lb. fat per year)	*Improved Yields (per cent)*	*Maintained Yields (per cent)*	*Lowered Yields (per cent)*
400 and over	6	20	74
380–399	9	27	64
360–379	16	31	53
340–359	24	37	39
320–339	34	39	27
300–319	52	32	16
Below 300	73	18	9

the lowest. However, on the average level and for the whole statement approximately one-third of the bulls fell into each class.

Consideration of positions in the normal distribution, regression to the mean or other matters of like nature is most important for characters which depend upon large numbers of simply additive genes. Conception of their inheritance is then in the nature of plus and minus arithmetical effects.

Very many characters of economic importance undoubtedly have this additive method of inheritance but it is not common to all. However, because it is so frequent and because mathematical processes can be applied unconditionally to data for characters controlled by it, the method of inheritance has largely influenced genetical research and thought. Nevertheless, it is not universal and, until methods of inheritance for all characters are determined such applications must always be made with reservations.

CHAPTER X

PHENOTYPIC SELECTION

It is evident from the foregoing that, had any of the early constructive breeders classified all the available animals of his chosen breed, either as individuals or with regard to their component polygenic characters, he would have had before him a series of normal distributions.

Discarding for the moment characters not of suitable nature, it is equally evident that, when he had made his selection and isolated it, he had a population which in many respects was no longer normally distributed. In doing so he had, as it were, pushed the curve off-centre. The problem before him was either to maintain it in that unstable position or to re-establish equilibrium in his new edifice.

On the one hand there was the tendency for progeny of his new herd to regress to the mean of the population from which it had been drawn. On the other, he had at his disposal, and used as methods of control, selection, and isolation with its implied inbreeding.

Imagine the aggregate of genes that was his herd as an inclined pillar of loosely-bound sand, ready to disintegrate and fall. Imagine this supported on the shoulders of a Colling, a Watson, or a Tomkins, while he pushed it upright and cemented it solidly by these two agents. This is the task each undertook with a degree of success made evident by time. These are the agents we must now consider.

PRINCIPLES OF SELECTION

The key to effective selection is to be found in the difference between the phenotype and the genotype of an animal (p. 89). In the beginning, the selection practised by all constructive breeders must have been phenotypic—based on the appearance of the animal. They had no other guide at that time. This method of selection is adequate for non-breeders or neuters such as wethers, steers, and the like. Also because every animal must be economically advantageous to maintain, whether it be a milch cow, a ewe, or a race-horse, it must have a phenotype associated with performance as of high milk-yield, fleece-weight, or a turn for speed. But the real value, as a breeding animal, is demonstrable only by the performance of its progeny. This depends upon its own genotype and that of its mate.

The phenotype is an ephemeral thing, a character of the individual, more often than not tied to the environment of the moment, a fleeting thing which dies with the animal and is lost for all time. But the genotype is an accidental combination of units that have their origin in antiquity and their future in eternity.

Genotypic selection is made when an animal is highly valued on the merits of its progeny. It is implicit in every instance of concentration upon the blood of a particular animal. How often that was an integral part of the methods of all the early breeders is made evident in previous sections. Thus, while their first selection must have been phenotypic, and while their adherence to type undoubtedly indicates phenotypic evaluations, there can be no doubt that genotypic selection was practised by them. The importance of the method justifies its discussion under the heading "Progeny-Testing".

REALISM IN SELECTION

Whether selection be phenotypic or genotypic, it implies the existence of criteria for desirability. These should be realistic and in accord with environmental forces whether primary or secondary.

Primary environmental forces are those of the geographical situation of animals. For example, climates are divided broadly into zones—the Arctic and sub-Arctic, the Temperate, and the Tropical and sub-Tropical,

Throughout geological time, each indigenous population has become adapted to its own zone. For the greater part, man's modification of these animals, and of the quasi-indigenous breeds and strains with which he worked, took place in temperate areas under relatively favourable conditions of available nutriment. An outstandingly important aspect of the early constructive work among British beef breeds was that, after selection of suitable kinds of animals from among the original cattle, they then, and probably for the first time, were provided with nutriment adequate to disclose their real differences.

Close association persists between the immediate environment, notably available nutriment, and particular characters in animals. This is especially so for those that have numerical values as pounds of wool or butterfat and pounds of beef, mutton, or lamb per head. Environmental control is as great today as it was in earlier times. Any massive bull, placid in its straw-covered stall, is the result of two forces—heredity and environment. The organism, which is the bull, must be in a state of stable equilibrium with the environmental forces. Here they are optimal, and the bull is massive and contented. Out on a northern cattle station, where the environment includes drought conditions, where heat is intense and there may be little or no food and sometimes no water, the identical twin of the placid bull and the environmental forces would not be in stable equilibrium. An increase of these adverse conditions would result in the bull's death, while amelioration would cause him to improve.

This raises a very important point; should conditions under which selection takes place be optimal or realistic? The first bull described, or a stud Merino ram under equally favourable conditions, would have the best possible chance to demonstrate all that was advantageous and inherent in his heredity. If similar environmental conditions were to persist, then selection among bulls or rams and their progeny in the same state would be proper. However, it is known that, when the sons of the bull or the ram go out to stations to beget grandsons and granddaughters of the favoured animals, their environment will be very different. An ability to survive under adverse conditions will be a criterion of excellence and capability of producing profitable carcasses or fleeces on minimal rather than on maximal levels of nutriment will be their most desirable characteristic. Because it is impracticable economically to interfere with the level of nutriment available under the natural conditions of station life, those of stud herds and flocks, which are flexible, should be adapted at least to simulate those that are unalterable. This may even involve developing entirely new breeds or strains of animals. However, the first step undoubtedly should be to eliminate the artificiality of nurse cows in beef studs. Female progeny of a bull whose mother could not provide sufficient milk to rear him adequately when conditions were ideal have not been selected for, and have very little chance of, nourishing their own progeny under much less favourable conditions.

Certain breeds and strains of domestic animals are peculiarly adapted to particular areas and purposes. These provide their true habitat and define their specific usefulness. Within them and for them, criteria of excellence, and so those of selection, should be equally specific. For example, the long soft coats of improved beef cattle are advantageous in cold countries. But this is not so within the tropics where, in addition to restricting loss of body heat, they afford more ready lodgment for ticks than short, sleek coats do. In one environment selection for, and in the other selection against, long soft coats would be equally logical.

An equally close association is necessary also with regard to selection and secondary environmental forces. All the principal breeds and strains of livestock are strictly utilitarian, and unless criteria for selection are compatible with local conditions and with the demands of available markets, stock raising will be unprofitable. When the English market requires heavy-weight baconers or short, thick and relatively light fat-lambs, these characters should be the criteria for selection and not those retained from earlier markets or the preconceived standards of breed societies.

Type, as defined by such societies, has played and can still play an important part in animal breeding. In some instances, however, it is unreal. This is particularly true in the scales of points laid down for dairy cattle. They allot points for appearance. In many cases they make provision for sire and progeny groups. But in no case do our breed classes require official figures to prove that the cattle are, in fact, the outstanding dairy animals they appear to be.

PHENOTYPIC SELECTION AND HEREDITY

The foregoing has reference to the general nature of animal populations. Further consideration will be facilitated by realizing that individuals and populations of animals are to be considered as aggregates of genes; that groups within these are the "flock sheep" of the earlier simile, with selection and isolation as the "wings" leading to "yards" where the aggregates will be confined and handled.

Gene groups within the aggregates may be fixed or unfixed. The former are responsible for all general characters that cause animals to be recognized as sheep, cattle, horses, or pigs. They cannot be interfered with by selection, but to some extent they can be unfixed by violent crossbreeding, such as species crosses, for example, horses and asses to give mules. It is the remaining unfixed genes that are amenable to treatment, but even these can only be effected within certain limits. Coincident pure lines for many characters cannot be expected (p. 76) nor will complex gene groups become homozygous readily. Even after close breeding and selection by Colling, which might have produced that state, Booth and Bates both were able to select and establish distinctive types from his strain of cattle. Had the methods reached their objective in the hands of Colling this would not have been possible without unfixing gene groups by crossbreeding. That they did not do.

The heredity of animals can only be effected at the time of mating; and the most common method of doing this is to select particular animals to leave progeny. Thus castration, spaying, and slaughter are similar processes. They eliminate animals assumed from their appearance to be repositories of disadvantageous genes. For the greater part, subjects for these operations are chosen either by their failure to satisfy a few essential criteria or for general failure with regard to many. When they are eliminated, the nature of the remaining gene aggregate, and so of the flock or herd, will depend upon the kind of character under consideration.

Characters Depending Upon a Small Number of Gene Pairs

If the character for which selection is being made depends upon a small number of genes, dominance can be assumed to be present. The retained portion of the flock will be homozygous if the recessive condition is desired, for example, red colour or horns in cattle, and pigmented wool in sheep. If the dominant character is being retained, for example, black colour, Hereford white face, or polledness in cattle, or non-pigmented wool in sheep, the selected portion of the herd or flock may be either homozygous or heterozygous. This would be so whether the selector knew the nature of the inheritance involved or was ignorant of it. Like to like matings will always reproduce simple recessive conditions, but they are not equally reliable for dominant characters. This has been explained in matings of heterozygous individuals (p. 87). They show the dominant character but, on occasion, will leave progeny with the recessive conditions. Like to like matings will give varying results also in specific heterozygous conditions such as roan in Shorthorns. These are more readily procured by mating the unlike parental types, in this instance red and white.

Polygenic Characters

When culling is for economic characters, for example, inferior make and shape, inferior fleece, and low milk-yield, the remainder, that is, the selected portion, must be referred to the curve of normal distribution (p. 97). All of these are polygenic characters and the culled portion will be approximately represented by the minus end of the curve (below the -1σ mark in Fig. 33). The remaining portion is not stable, however, because most of the gene groups will probably be in a heterozygous state. Individuals, therefore, will be able to reproduce all types found in the original flock. This will happen if the selected females are mated with males selected from the same flock by the same criteria. Their progeny will show about the same range from very bad to very good with a comparable average and no progress will be made.

It is customary to use males that are more highly selected than the females with which they are mated. Thus there may and probably would be a slight movement of the average towards the higher or plus side of the original distribution. The extent of this movement will be controlled largely by the relative excellence of the sires used. They, depending on their breeding worth, would introduce more or less advantageous genes. When the difference between the males and females is not great, upward movement, that is, improvement, will be slight. If the bias given by selection be reduced or the rating of the males be lowered, so little improvement will take place that distribution of characters in the next generation will be similar to that of the original flock.

When a maximal and long-continued effort is made at these two points—selection and superior males—the character of the gene aggregates may and probably will change. Very gradually, generation by generation, individual groups will approach a homozygous state if and when related males are used. Selection will have eliminated undesirable individuals and genes and a slow improvement will be obvious. These principles have particular reference to the general average of fleece-weights over a period of years within the Australian wool clip. Their association with that statistic will be discussed in a section dealing with breeding policies for the wool industry.

It must be realized, however, that there are definite limits to phenotypic selection as a method of breed improvement. Lewis (1942) made a study of 5180 "commended" and "highly commended" Island Jersey cattle and their parents. It had reference to seventy years of selection for conformation in these cattle. He found that the proportion of "highly commended" progeny to "highly commended" parents showed a considerable increase—23 to 43 per cent—in the early years of selection. After that advance, although fluctuations occurred, there was no consistent increase. The method was unable to increase homozygosity for genes which influence "highly commended" conformation and produced its maximal effect when first applied.

Classification as "highly commended" has reference to the appearance of the animal. Performance, however, is also phenotypic, and selection for it gives results of the same nature.

This is made clear by figures, over sixteen years, for the average production by all pure-bred cows under official test in Western Australia. Table IX presents figures for the yearly averages of production by these cows. It will be seen that the method has made very little improvement in the averages for milk and butterfat. It must be remembered also that feeding standards would have been improving all the time.

SELECTING UNLIKE MATES

Sometimes the selected portion of the flock or herd is mated with dissimilar males to give compensatory matings, strong points in the males offsetting weak points in the females. For example, an owner, after culling his ewe weaners, may split the remainder two ways. There may be finer or stronger wool groups and more or less "developed" groups. These would be mated, the fine-

TABLE IX

AVERAGE PRODUCTION OF PURE-BRED COWS UNDER OFFICIAL TEST DURING 16 YEARS

Year	*1924*	*1925*	*1926*	*1927*	*1928*	*1929*	*1930*	*1931*	*1932*	*1933*	*1934*	*1935*	*1936*	*1937*	*1938*	*1939*
Gallons of milk ..	600	652	624	602	592	629	636	643	696	664	720	682	681	685	664	681
Pounds of fat ..	319	309	312	291	281	295	295	302	319	309	334	327	320	309	307	310

TABLE X

SPINNING COUNTS AMONG FIRST-CROSS FLEECES

Ram no.	*Sex of progeny*	*Counts*													*Totals*
		46/50	*50/56*	*50/56/58*	*54*	*54/56*	*56*	*56/58*	*58*	*58/60*	*58/60/64*	*60*	*60/64*	*64*	
330	Ewes	1	1			1	5	6	5						19
	Wethers			1			2								3
	Rams		1				2								3/25
990	Ewes									3		1	2		6
	Wethers							1	2	2		2	2		9
	Rams							1							1/16
993	Ewes							8	1	8	1	5	4	1	28
	Wethers							1	2	4		2	6		15
	Rams						1	2	4					1	8/51
994	Ewes				1				4	5		4	1		15
	Wethers										1				1
	Rams							3	3	3		2			11/27
TOTALS		1	2	1	1	1	10	22	21	25	2	16	15	2	119

wool ewes with strong-wool rams, the more developed with plain-bodied rams and the reverse.

Each parental type subscribes equally to the progeny so that gene groups responsible for individual characters establish new combinations. The outcome of this is uncertain. It will depend upon the degree of dominance involved and upon the state of gene groups responsible for the character, that is, whether they are more or less homozygous in individual parents.

Table X presents relevant data. These are too few to permit conclusions, but they illustrate points being made. The figures show the number of first-cross long-wool x Merino sheep, with spinning counts of various kinds. These were determined by an experienced appraiser during shearing. No. 330 is a Border Leicester ram mated with strongly developed Merino ewes. Numbers 990, 993, and 994 are very developed Merino rams mated with Border Leicester ewes. The first-cross sheep are mixed with regard to sex which is shown.

It will be seen that the totals disclose a distribution comparable with those already presented (Tables IV and V). However, in this instance it appears to be "skewed". Broadly, most of the fleeces, 66 per cent, fall between 56 and 58/60 which is about intermediate to the two parental types. This and the general picture conform to polygenic types of inheritance. Individual rams appear to behave differently, however. Twenty per cent of progeny by No. 330 were below this range, while 37 per cent of all the progeny of Merino rams were above it. This could be explained by suggesting that No. 330 had been bred from Border Leicester ancestors so strongly selected for coarse wool that the appropriate gene group carried by No. 330 was more or less homozygous for the condition. The converse of the suggestion could be advanced for all the Merino rams. However, their progeny were also different. Nos 990, 993, and 994 had respectively 44, 37, and 26 per cent of their progeny above the 58/60 range (the two sheep in the 58/60/64 class are discarded). These figures indicate that the genetic constitution of the Merino rams was different and suggest that No. 990 was more nearly homozygous for fine wool than No. 994. As stated previously, the data cannot be regarded as conclusive, but are suggestive and adequate for illustration.

An hypothesis which has been put forward to explain such observations suggests multiple additive genes are either plus or minus. Any parent can have all plus pairs, all minus pairs, or some plus and some minus genes paired. The contribution of each parent to its offspring having been subjected to reduction division, it will consist of a larger or smaller number of single plus or minus genes depending upon the parent's heredity. All of these single genes in the two gametes become paired in the zygote and balanced at a point related to the numerical strength of either kind of gene. That balance controls the phenotype or appearance of the offspring. On this hypothesis, if coarse fibres were controlled by plus genes and fineness by minus genes then the ram B.330 would have contributed more plus genes than the Border ewes and the different Merino rams would have contributed different numbers of minus genes.

CROSSBREEDING AND HYBRIDIZATION

When unlike mates are sufficiently dissimilar to be identified as from different breeds, then crossbreeding takes place and the progeny are called crossbreds. Similarly when mates are so dissimilar that they come from different species, then hybridization takes place and the progeny are hybrids.

The principle underlying use of two such dissimilar mates is that the change desired in the progeny is beyond the range of possibilities for genes within the aggregates constituting the breeds. Imagine particular recognized breeds as quiet landscapes in which the ants, beetles, birds, and other normal inhabitants are the genes, all busy after the manner of their kind. Crossbreeding is the 25-pound shell that shatters the position. When the smoke clears away, the magpie finds it has a mudlark for a mate and the ant a spider in its nest—that is what happens to the genes.

Crossbreeding and hybridization unfix the maximal number of genes, add others, and give rise to a very high proportion of heterozygous groups. This has several consequences. It provides the greatest number of possibilities for selection, because all the heterozygous forms will give rise to progeny having at least minor variations. It provides such a degree of vigour and vitality by unfixing, adding, and recombining gene groups, that the phenomenon is referred to as "hybrid vigour". This is greatest when the parents are most genetically diverse, for example, the inter-species cross of horses and asses to give mules. Crossbreeding introduces altogether new genes to the previous aggregates. For example, Merino ewes are relatively poor milkers and are rangy in build. Their first-cross progeny by long-wool sires receive new genes from that breed and are much better milkers, larger and more nearly square, than their Merino mothers. The Southdown ram, however, is still closer to the ideal, short, thick, square-jointed carcass of light weight desired by oversea markets. Accordingly, a third cross is made and another series of genes is added to the already modified aggregate. The second addition makes the superior export fat-lamb possible.

Similar pooling of resources takes place, for example, with regard to resistance to ecto-parasites and to "type". Pure-bred Zebus are highly tick-repellent, but do not have the shape characteristic of British-bred cattle. These, on the other hand, are susceptible to some ecto-parasites, notably ticks. The first cross of the two breeds gives progeny nearly as tick-repellent as the pure Zebu and not unlike a British beast in make and shape. It is about intermediate, but bears somewhat towards Zebu characteristics. Repellence increases, and similarity to British cattle decreases, when Zebus are the recurrent parents in further matings. Corresponding changes take place if the first-cross cattle are mated and their progeny remated with British cattle.

Crosses between animals of different strains within a breed and those between breeds or species differ only in degree. The fundamental difference lies in the dissimilarity of the genes they introduce to the new aggregate. It will be remembered that all domestic breeds and strains have benefited by this procedure. Originally they were in a markedly crossbred state, a condition of unfixed genes. It will be shown how isolation and so close breeding, which was the next step, reduced the unfixed state. However, every instance of outcrossing, whether the slight degree practised by Bates and Booth or the more violent procedure of Cruickshank, was similar in principle. McCombie formed new aggregates by unfixing genes segregated and fixed by the methods of Fullerton and Watson. Hereford breeders fixed gene groups within variously coloured strains. When these were amalgamated reciprocally, a new source of variation arose from the unfixed condition of the genes and new combinations thus became possible. This is fundamental to all crossbreeding.

SELECTION BY PERFORMANCE

Whether it be on the race-track, on the wool floor, or in the milking shed, performance is a character of the individual. It is as truly phenotypic as colour or any other attribute. Performance is a very discriminating aid to selection and, because it is more realistic than anticipations of merit based on make and shape, it is of greater value. Thus records of performance collected in Turf Registers, Official Herd-Test Records, Registers of Merit, and similar files are very often used as guides. Matings may be between similarly good performers, as with horses, or, as with dairy cows, between a good performer and the son of a good performer, in which case they are like to like matings and behave as such for polygenic characters. Sometimes matings are compensatory, as when "stayers" are mated with "sprinters". These characters also can be expected to behave normally as for unlike mates.

The most important point to observe in respect to the

use of records of performance as bases of selection is the preponderant influence of environmental forces upon most outstanding performers. Thus, as indications of hereditary characteristics of the animal in question, they may be misleading. This is important in assessing dairy cattle, particularly their occasional performances such as short-term production tests. To overcome the difficulty herd-testing associations frequently insist upon normal conditions; specify the number of daily milkings, the duration of lactation, the interval between calvings, and the repeated testing of all the animals in the herd. Such data provide lifetime averages which are most reliable. Other methods compare the production record of an individual with the average of her herd-mates' production in the same period. Performances on the race-track can generally be regarded as being made under optimal conditions of environment and practically constant in this respect.

A safe plan is to discount all outstanding performances and to classify animals that make them as "above the average", having reference to the normal distribution (Fig. 33). Thus butterfat records of 800 lb. in 273 days, in this respect, may be regarded as comparable with those of 600 lb. when the breed average is 400 lb. and the highest point in the range ($+1\sigma$ Fig. 33) of the majority is in the order of 450 lb. The suggestion is more realistic than hopeful but, when associated with matings of phenotypically like to like animals, does not restrict the probability of uplift in the general average.

GRADING-UP BY SELECTIVE MATINGS

When a line of scrub cattle or of inferior sheep is mated for generation after generation with pure-bred bulls or rams of adequate merit, the flock or herd is greatly improved. The lengthy nature of the process differentiates it from crossbreeding. However, at the same time, food-supply, management, including culling, and other environmental factors appropriate for the improvement must be provided.

Culling affects the distribution of the flock or herd as

discussed previously. Within the limits of phenotypic selection it eliminates disadvantageous genes. Introduction to the culled herd of a dissimilar sire, which is much more meritorious and assumed to carry correspondingly advantageous genes, adds these to those remaining from the original aggregate. The distribution of the next generation, therefore, is erected upon a range of possibilities different from the original, and the mean moves toward the plus side of that range (Fig. 33). There will still be a tail on the minus side, however, which must be wiped out. Progressively the mean or average of the group moves upward. It will always be somewhat lower than the value attributed to the introduced sires, because regression to the mean of the original population still is an operating force, and the introduced sire can only affect the position in proportion to the extent that his excellence is hereditary.

The method is an extreme case of compensatory matings between dissimilar parents. Although it frequently occurs, an animal from a new breed is not necessary to provide the dissimilarity. The superior males can be of the same breed, for example, a superior Hereford bull in a run-out Hereford herd or superior Merino rams on a run-down Merino flock. The principle involved in using a new breed is that with it a maximal number of new and advantageous genes is introduced. Although considerably restricted at present by economical and practical limitations, artificial insemination is a particularly suitable method for grading-up processes. The reduced number of superior males needed permits the purchase of outstanding animals with the same appropriation of funds. With them, more speedy improvement is possible, particularly if their merit is judged by preliminary progeny-testing.

CHAPTER XI

GENOTYPIC SELECTION: ISOLATION

GENOTYPIC SELECTION

"By their fruits ye shall know them"—that epitomizes genotypic selection. While phenotypic selection assumes, genotypic selection assures that preferred animals will have advantageous genes. It discovers parents that leave specific kinds of progeny, and eliminates the others. It is used to control characters due to many genes when environmental forces are powerful; particularly when, as in milk-yield or egg-production, the character is restricted to one sex. However, it is not infallible, because performances of progeny are themselves influenced by environment, and the heredity of progeny is subscribed to equally by the other parent. Also, offspring have fortuitous gene combinations inseparable from Mendelian inheritance and so may provide misleading information. Nevertheless, it is the soundest procedure for classifying breeding stock.

In previous sections the sampling nature of Mendelian inheritance has been emphasized. It is the principle which underlies progeny-testing: repeated samples are drawn from the aggregates of genes that are the heritable material of parents. Within limits the greater the number of samples, the more reliable will the estimate be. The samples are appraised from characterization of progeny. Thus the test becomes more efficient as the number of progeny increases, until the law of diminishing returns begins to operate. The number necessary is largely influenced by the degree of variation within the samples, that is, by the magnitude of differences between individual progeny. If a cask of numbered marbles were agitated between draws, and each draw disclosed only marbles marked with the number two, a very few handfuls would provide a satisfactory test of its contents. On the other hand, if each draw were to disclose a wide range of numbers, many more draws would be necessary to estimate adequately the contents of the cask. Before the appropriate number of samples, that is, progeny, can be determined, it is therefore necessary to discover the degree of variability within them. There are mathematical expressions for this, the most common being the coefficient of variability. It is the standard deviation (p. 97) expressed as a percentage of the mean. It is beyond the scope of this discussion, but when determined it can be used in statistical tables to indicate the necessary number of replicates, that is, samples or progeny. The point has been raised to show the advantage of measuring the character for which the animal is being tested. Calculations can only be made from figures. Qualifications like good, bad, fair, must be changed to point scores for this purpose, and should not be used if the character can be measured, as fleece-weight, or pounds of butterfat.

Measurement of characters, although advisable, is not essential. Experienced judges can differentiate between the merits of several groups of progeny by examination and appraisal. Thus they can determine which, in their opinion, is the best of several sires. However, memories are short and difficulties arise when it becomes necessary to determine improvement or differences between animals when one is dead or the progeny groups are far apart, either geographically or in time. Suitable systems for recording evaluations can be invented to overcome these disadvantages. For example, at one Merino stud, from among the get of several sires, the number of young rams classed to the various stud-grades is used to compare the sires. The grades are arbitrarily valued at say from £100 to £5. The sires having been given equal opportunities, a figure for each is determined from the sum of the amounts representing the number and value of the sons of each in the several grades.

Among Thoroughbred horses, results of an extensive progeny-test are disclosed periodically in lists of successful sires. Their principal disability is that they do not record the performances of poor progeny. For beef cattle the best criterion by which to evaluate sires has been shown to be the rate of gain in body-weight per unit of food consumed by their progeny. The best sire is one whose progeny consume least food for a standard unit of gain in body-weight. The Danish criteria for pigs were those of minimal food consumption for maximal carcass quality among their progeny. To test any sire adequately, however, it is necessary to appraise all or a large random sample of his progeny. To consider only the best would be misleading.

Bull Indices

As early as 1913 Hansson of Oslo University suggested that a rating of dairy sires for fat percentage could be obtained by placing the daughters half-way between sire and dam. That this is approximately so is shown by the data of Wriedt for crosses between Red Danes and Jerseys (p. 99). Hansson's suggestion was published in the Swedish language and was overlooked by most animal breeders. However, a number of other attempts to appraise sires by the production of their daughters were taking place. Prentice (1935) states and discusses eight of these, but only the two most widely used are considered here.

Indices are designed to indicate a figure likely to represent production by the bull, if that were possible. The number of daughters sufficient to give a valid estimate is the first consideration. Edwards (1932) has prepared the following table which shows how accuracy of the estimate increases with the number of daughters.

The figures for pounds of milk indicate the difference which must exist between yields of a given number of daughters of two bulls being compared before one can be said to be significantly better than the other. For example, in the case of two bulls each with one daughter, the difference in the daughter yields would need to be

5521·5 lb. of milk. It will be observed in the table that decrease in the margin of difference required takes place most rapidly in the first few daughters. Possibly six daughters are minimal but every daughter's record additional to the list increases accuracy of the estimate and all daughters must be unselected.

TABLE XI
EFFECT OF NUMBER OF DAUGHTERS ON ACCURACY OF THE PROGENY TEST (AFTER EDWARDS)

Number of Daughters	*Significant Difference lb. milk*
1	5521·5
2	3904·8
3	3187·8
4	2760·9
5	2469·3
6	2253·5
7	2086·4
10	1746·3
13	1531·2
15	1425·6

Having established the minimal number of daughters' records it is necessary also to know those of their dams. Either record may have been made at a different age and under a variety of conditions. Correction figures have been evolved for age, number of milkings, seasons of calving, and the like, but adjustment for age is the most common and necessary correction. Factors for this, abstracted from the Agricultural Gazette of New South Wales and approved by all Australian breed societies are as follows. Age is taken from the beginning of the lactation period.

Junior 2-year-old (under 2 years and 6 months)	add 33 per cent
Senior 2-year-old (between 2 years and 6 months and 3 years)	add 25 per cent
Junior 3-year-old	add 20 per cent
Senior 3-year-old	add 12 per cent
Junior 4-year-old	add 7 per cent
Senior 4-year-old	add 5 per cent

With these data it is possible to calculate an index for the bull. The first is known as the equal parent index. It is based upon an assumption similar to that stated earlier for butterfat; namely, that, on the average, production of daughters is half-way between the parental levels. The formula is: sire's index = twice the average production of his daughters minus the average production of their dams, or S.I.=2 Dr.—D.

Study of a large volume of data and of Gowen's cross-breeding experiments to determine inheritance of milk and butterfat convinced Dr Goodale of Mount Hope Farm, U.S.A., that allowance should be made for partial dominance of high milk-yield and recessiveness of high butterfat percentage. Accordingly, he evolved two indices to be used either when the daughters exceeded their dams' production or when the dams' was greater than that of their daughters. Calculation of these indices is more complicated than for the equal parent index which is used as a commercial form for classifying bulls of the American Cattle Club, although the club is supported strongly by Mount Hope.

A third form of index was suggested by Gifford as simply, Sire's index = average production of daughters. Edwards, who considered accuracy of all three found this third index yielded smallest deviations from expected levels of production when they were compared with actual figures. However, because the first is most frequently used, examples of calculating that index for milk and butterfat are given as follows: (i) daughters' average mature equivalent production of milk, 10,000 lb.; dams' average mature equivalent production of milk, 8000 lb.; sire's index is 20,000-8000 or 12,000 lb. milk; (ii) daughters, 8000 lb.; dams, 10,000 lb.; sire, 6000 lb.; (iii) daughters' average butterfat per cent, 4·5; dams' average butterfat per cent, 4; the bull is represented as having (2×4·5)—4, that is, he is a 5 per cent bull.

If a bull has 7 daughters, whose average production is 10,000 lb. of 4 per cent milk, from dams whose average production is 8000 lb. of 3·6 per cent milk, the index of the bull would be stated as follows: 7 pairs: 12,000 lb. milk; 4·4 per cent, that is, 528 lb. butterfat.

No index is an infallible guide to the transmitting power of a bull, because so much depends upon the nature of the dams. This is particularly true when the same bull is used in different breeds. Some other indices attempt corrections for this and other factors but their calculation becomes complicated.

It should also be realized that any form of averaging gives rise to a degree of smoothing of data which frequently conceals essential peculiarities within them. For example, a bull may beget progeny equal numbers of which are above or below the production levels of their dams. However, if the excesses were numerically greater than the deficits, the bull's index may be similar to that of another bull which uniformly begets progeny equal to, or only slightly in excess of, the dams. Nevertheless, the second would be the better bull.

For such reasons many researchers will not use any bull index. They prefer to state the facts graphically. Thus sire surveys in New Zealand report results as associated curves which are two lines drawn from points representing production of the dams and of their daughters. In America it is customary to present the data as arrow charts. Production of a dam is spotted as a point on a graph and from that point a line is drawn, up or down, on the same scale to the point representing her daughter's production, which is marked by an arrow head. Every dam-daughter pair is plotted separately and the whole series for a bull's progeny constitutes his arrow chart. The bulls are then classified by the nature of their charts as (*a*) proven for all levels of dams, (*b*) proven for dams below a certain level at which the bull breaks down, (*c*) proven only for dams up to a certain level of production, or (*d*) as so erratic in his test that, although he may leave some daughters markedly higher than their dams, he is unreliable and so not proven.

Progeny-testing for Recessive Conditions

There is another type of progeny test designed specifically to examine the genotype of a sire for recessive conditions. Heterozygous animals can be expected in any normal population. Occasionally the recessive genes they carry are responsible for the death or disability of their progeny from particular mates. The test provides sires with mates likely to have latent conditions. The most likely of all are his own daughters. If a ram is heterozygous he must pass recessive genes to some of his progeny. Those genes will have an opportunity to complete their pairing and to disclose the condition when the ram mates with a sufficient number of his equally heterozygous daughters.

Such a condition and its repercussions were discussed by the author in the *Journal of the Council for Scientific and Industrial Research*. The condition in question was shortened lower jaw. In sheep with this fault, the incisors oppose the palate instead of the dental pad, and the apposition takes place approximately three-quarters of an inch posterior to the normal position. A Merino ram, M3, was depastured with an experimental group of Merino ewes, and with his progeny from them. While running with the flock he mated with his own daughters, bred within the group, and later with his own granddaughters.

By January 1942 twelve of his daughters had borne progeny to him. One had borne three lambs, four had borne two, and seven had each borne one lamb. In 1940 the original ram (M3) was withdrawn temporarily and was replaced by one of his sons. This son, A118, left one ram lamb out of a ewe by his father.

Typical pedigrees from some of these matings are shown in Fig. 37 (*a*) and (*b*).

The pedigree outlined in Fig. 37 (*a*) shows A91 as a daughter of M3, and the mating between father and daughter with the product, B376. The lower jaw of this lamb, B376, was so abnormal that it was impossible to rear him.

The pedigree outlined in Fig. 37 (*b*) shows the product B648, of mating a son and a daughter of M3. This lamb also showed the abnormal condition of the jaw. However, he has been stall-fed throughout, with good prospects or rearing him for further experimental matings. The ewe A105 bore three other lambs to M3—all were normal. The ewe A91 has borne only the lamb B376.

There were two kinds of ewes in the original group, those with the prefix letter H, and those with the prefix letter K. The history of these sheep is well known. Until

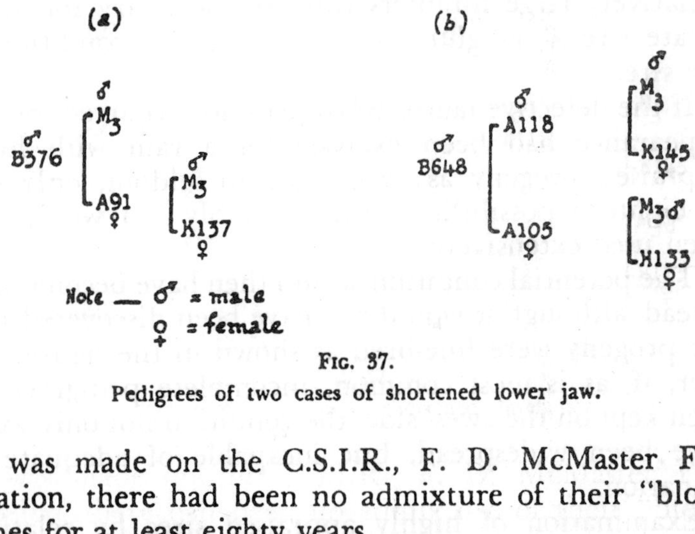

Fig. 37.
Pedigrees of two cases of shortened lower jaw.

it was made on the C.S.I.R., F. D. McMaster Field Station, there had been no admixture of their "blood" lines for at least eighty years.

Shortened lower jaw is a recessive condition discovered by the inbreeding. Further, because the only relationship common to the ram lambs B376 and B648 is that to M3, unless it is very widespread, the condition must have been inherited through him. It is of interest to sheep breeders who are progeny-testing sires, and of particular interest to those who, by artificial insemination, are extensively using sires appraised only by the relative excellence of progeny from ewes not closely related to the rams.

The ram M3, when mated with ewes other than his own daughters, left seventy-eight lambs, all with normal jaws. His excellence as a sire would have been appraised from an examination of the wool and other characters of these lambs, and it would have governed his future use. If, as a stud ram, he were appraised highly, probably he would have been used extensively by "hand service" or by artificial insemination.

It was not until he was mated with twelve of his own daughters that the condition was discovered, and then it only occurred once among eighteen lambs from them. It is emphasized that M3 begat three normal lambs from the daughter A105, which later bore the abnormal B648. Relatively large numbers thus are necessary for appropriate sire x daughter tests for recessive conditions in the sire.

If the defective lamb, B376, had not occurred, or if its appearance had been excused for a ram with highly appraised progeny as happening so seldom, only once in eighteen possibilities, then possibly M3 would have been used extensively.

The potential condition would then have become widespread, although it would not have been discovered until the progeny were line-bred as shown in the figure. Further, if, as is most common, incomplete pedigrees had been kept on the ewes' side, the condition not only would have been widespread, but incapable of adequate explanation.

Examination of highly appraised sires by relatively numerous sire x daughter matings is essential before such sires can be used extensively with safety.

Selection by Pedigree

Growth of selective registration brought with it the conception that names in a pedigree were themselves hallmarks of excellence. It is clearly shown in the early sections that this fetish was perpetuated by the importance attached to names of particular animals. Thus there were the Queen Mother family, the Duchess family, the Trojan Ericas, and a number of others in which every pedigree traced eventually to a famous animal generally in its bottom, or female line (Fig. 12). Bruce Lowe contributed to it by stating that existing Thoroughbred horses were markedly influenced by the "blood" of one or other of some forty remote female ancestors, of which approximately twenty-five lines still remain. Thus pedigrees have been depreciated in their usefulness. Nevertheless, with certain reservations, they can be and are very valuable documents, and their proper use may be a type of genotypic selection. However, while pedigrees look backward from a potential parent, progeny-testing looks forward for its criteria of excellence.

Essential features of a pedigree for breeding purposes are complete records of the nature and relative merit of all recent ancestors and also of closely related individuals. As a statement of names and without this additional information or value as an index to it, a pedigree ceases to be a valuable document and becomes only the trade mark of a breed society.

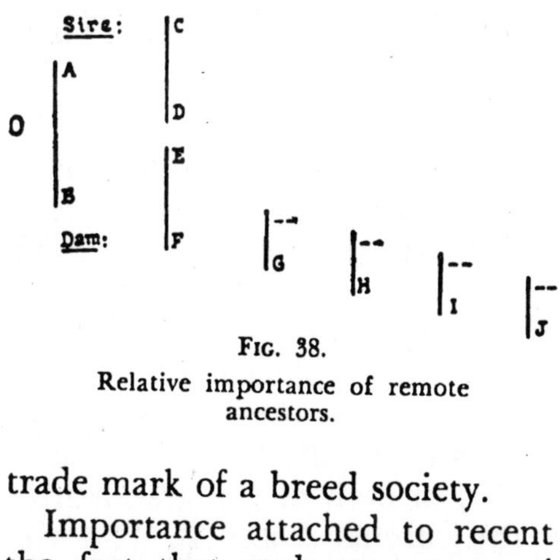

FIG. 38.
Relative importance of remote ancestors.

Importance attached to recent ancestors arises from the fact that each parent provides half the heredity of its progeny. It follows, therefore, that a grandparent, contributing half the heredity of the sire or dam, will supply only half of that half, or one-quarter, to their progeny. Thus the contribution of each ancestor is progressively reduced by one-half multiplied by itself as many times as there are intervening generations. For example, in Fig. 38, A or B contributes half the inheritance of their offspring O. C or D contributes half that of A and so one-quarter of that of O. E or F as a maternal grandparent is equally important, and supplies one-quarter of the heredity of O. Similarly G supplies one-half of one-quarter, or one-eighth; H, one-half of one-eighth, or one-sixteenth; and so on until J supplies one-half of one-thirty-second, that is, one-sixty-fourth. These calculations could be continued almost indefinitely among registered livestock.

It is obvious that hereditary characterization of O is more largely dependent upon contributions by ancestors prior to, say, G in the third generation back, than from those more remote, for example, J. It should be realized that the conception of equally distributed contributions can have reference only to genes which produce additive effects without dominance, linkage, or other complicating actions. However, when those genes are excluded, the remainder, which are additive, probably includes most of those responsible for characters of economic value. The conception, therefore, is tenable.

The principle underlying insistence on information with regard to ancestors and relatives is based upon the fact that all are samples of the aggregate of genes present in any generation. Broadly, the genetic material fundamental to the sixty-four ancestors in the generation of J is that found in any other generation. It was possible, however, for J to provide innumerable samples, any one of which might be different from any other in its particular gene combinations. This is equally true of all in the generations of I, H, and G, of each of the grandparents C, D, E, and F, and of the parents A and B. The Mendelian conception of variation explains and anticipates this sampling. A is not the only possibility from mating C and D. A or B or any one of their full brothers or sisters is a sample from many possible combinations of the gene aggregates or genotypes of its parents. It follows, therefore, that any appraisal of the hereditary potentialities of O becomes more credible as the volume of evidence increases with regard to the merits of other samples drawn from the same aggregate. If, for example, O were a sealed case of apples containing fruit secured from each preceding case, in the proportions discussed, and it were possible to tip out and examine the contents of cases C, D, E, and F, there would be no uncertainty with regard to the quality of the fruit in the case O. The two instances are completely analogous within the limits stated above for restricted mechanisms of heredity.

Relationship

Evaluations in the foregoing show why similarity between animals results broadly from their relationship one to another. Related animals have genes, and so characters, derived from similar sources. These become similar if the animals are sufficiently closely related, for example, full brothers and sisters and the extreme case of identical twins.

In human populations these relationships are seldom complicated, and everybody is familiar with what is meant by the expressions half-brother, half-sister, first or second cousin and so on. They convey to the mind an unconscious expectation of similarity. Among animals, however, these simple relationships do not describe all possibilities. Many would tax equally man's ingenuity and his vocabulary for adequate description. To overcome this, relationships have been expressed numerically. There are two principal methods, the percentage of blood and the coefficient of relationship evolved by Wright (1921). The first, although it does not include collateral relationship, is very simple to determine. The second requires more extensive calculation but takes cognizance of all relations and is particularly useful to determine the relation one animal bears to another that is not a direct ancestor. However, because its determination is more complex, description of the method has been relegated to the appendix together with that of Wright's coefficient of inbreeding.

Calculation of percentage of blood relationship requires only summation of fractions already shown as measuring the contribution of simply additive genes by each ancestor to offspring in a particular generation. Thus when the name of an animal recurs frequently, the percentage of blood relationship to it is the sum of the appropriate fractions. Fig. 39 shows a pedigree in which each letter represents an ancestor, some of which are recurrent. The percentages below each generation show the contribution of animals in that generation to the heredity of their descendant, O. The percentage of blood relationship of O to C is 25 + 25, or 50 per cent; to F or G it is 12·5 + 12·5, or 25 per cent; to L, M, N, or P it is twice

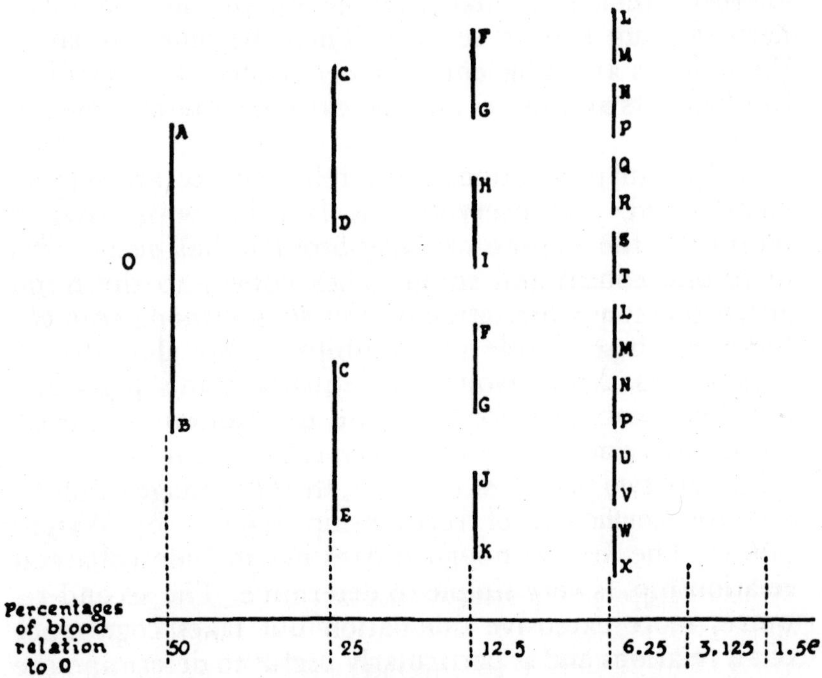

FIG. 39.

Pedigree to illustrate "percentage of blood" relationships.

6·25 or 12·5 per cent, while to any other animal it is the percentage appearing below each generation.

Percentages so determined do not state any outstanding biological feature of the animal under consideration. They merely translate into figures what is implied by stating, "This is the second cousin of that." However, they are convenient for the purpose, and are used fairly generally in the United States.

ISOLATION

The second important agent employed by early constructive breeders was isolation of their herds. They did not originate the method, although Bakewell apparently was at least one of the first to employ it deliberately. Isolation has always been an important cause of uniformity in livestock occurring naturally on islands and in secluded districts. Probably most of the sheep and cattle mentioned by Youatt as typical of parts of the British Isles owed their uniformity to isolation as a natural process.

It produces this effect by confining all sampling, and so recombinations of genes, to the aggregate of the particular flock or herd. At first these aggregates would contain many gene groups that were heterozygous. They would be much more complicated, and each animal would represent many of them, but the original state can be illustrated by reference to the relatively simple heterozygous types discussed earlier (Table III). It was shown that, because chance controlled the sampling, the two parental types would give rise to two different types in addition to reproducing themselves. Under natural conditions, some of the divergent types would perish or be unthrifty while others would survive. This is the great explanatory contribution made by Mendelian ideas of inheritance to the problem of the origin of species. When man added selection to these natural tendencies, certain types would be preferred and eventually form the larger part of the herds. These animals, because they were isolated, could only mate with one another. Recombinations of genes resulting from such matings would all be restricted by a fixed range, and, accompanied by eliminating selection, would continually increase the proportion of homozygous animals in the population until the position was stabilized. The change would be progressive, and uniformity in the flock or herd would also increase.

Breeding groups may be isolated deliberately with similar results. This has been done many times since

Bakewell founded his Dishley Society. Established policies of breed societies and registration by them only of progeny from registered parents isolates groups of animals as breeds. Restriction of sampling to an aggregate of genes defined in this way gives rise to the same slow movement toward a higher proportion of homozygosis in the breed, as was evident in groups isolated naturally. At the same time, specifications by the societies of scales of points for type, and occasionally inspections by officers of the societies, ensure the elimination of "off types" and so of undesirable gene combinations that occur as a result of sampling.

If no preferment were exercised for particular strains or families, the rate of this movement within the whole breed would be comparable with that discussed for "like to like" matings in smaller groups (p. 112). However, within registered "pure breeds" predeliction is expressed for particular animals. When it is restricted to that of one person or to those who control only a relatively small part of the breed, it will cause the part to be isolated and to become different from the remainder. But, if the predeliction were general, then the breed would move as a whole.

There are these three major possibilities for animal groups that are registered or otherwise isolated. The first is that every animal, because registered, should be regarded with equal favour. The reduction of heterozygosis then would depend upon the number of males in use and would be comparable with that of animals isolated in a natural state. Wright estimates the reduction per generation under these conditions to be $\frac{1}{8M} + \frac{1}{8F}$, where M and F respectively are the numbers of males and females in use for breeding. The second possibility is that the breed will be split into a number of sub-groups which will move independently toward uniformity and homozygosis. This is the position of Merinos in Australia. Our parental studs that are "closed to outside blood" form such groups. Each of these has its own favoured animals, and the reduction of heterozygosis and with it increasing uniformity depends upon the breeding methods employed within the stud flocks. The third possibility arises when particular animals are widely favoured and much sought after. Favourite in Shorthorns, Sovereign or Sir David in Herefords, Watson's Jocks in Aberdeen-Angus, Eclipse in Thoroughbreds, and Prince of Wales or Darnley in Clydesdales, are all instances of such animals. When this takes place the breed moves forward with greater speed than in the first possibility and with greater uniformity than in the second.

Genctical examinations of registered breeds disclose these kinds of isolation. They are able to assess as a percentage the proportion of genes remaining in an unfixed or heterozygous state. Such estimates are seldom more than seven per cent for fixation, or homozygosis. Thus an increase in homozygosis of from zero to five per cent was estimated for Ayrshire cattle between 1877 and 1927. Lush (1939) has made many such examinations. Most of them have had reference to British breeds and to their subsequent history in the United States. He stated that "most pure breeding reduces existing heterozygosis in each generation by 0·5 per cent". He defines a generation (Lush 1937) as being "the average age of parents when their offspring are born", and states this to be four and a half to five years for beef cattle, four to four and a half years for dairy cattle, and four to four and a half years for sheep. He continues, "The reduction proceeds mostly from the efforts of so many breeders to have their herds headed by sons and grandsons of very few concurrently famous sires. . . . There is only a faint tendency (in America) for families within a breed to be any more distinct than they would be if the breed were truly a random breeding population." That is as in the first possibility above or in the group isolated on an island. "The close similarity," he states, "of the finding from so many different kinds of breeds makes it appear that these conditions are almost universal."

Lush's estimate of a generation, as defined, is approximately four to five years, and his figure for reduction of heterozygosis in each generation is 0·5 per cent. At the rate disclosed it would take about 56 years of normal procedure for breeding registered stud stock to reduce heterozygosis in the breed by as great a proportion as one mating of first cousins, that is, 6·25 per cent. When it is realized also that one generation of brother and sister mating raises the percentage of inbreeding to 25 per cent, the rate of increase in gene fixation indicated by the above figures for registered breeds becomes very small. It is because of this that those commonly referred to as pure breeds have been differentiated from genetically pure breeds. Further, these figures throw considerable light upon the possibility, frequently discussed by botanists, of fixing pure lines of animals (see also p. 109). The variety of types to be seen in any so-called pure breed bears testimony to the reasonableness of low estimates for the proportion of fixed to unfixed genes remaining in their aggregates. Uniformity would result from maximal fixation and so a high calculated percentage of inbreeding. For example, continued full brother and sister matings for 16 generations would result in 95 per cent as a coefficient of inbreeding. There would be a high degree of uniformity from about the tenth generation onward.

Inbreeding

Early sections disclosed how frequently favoured animals were mated with their own close relatives. Indeed in-and-in breeding was characteristic of the early constructive period and it is still a very common practice. The method restricts considerably the range of possible recombinations of genes and, by repeatedly sampling the aggregate of a particular individual, causes all samples to be relatively similar. There is no new principle involved. The method is one of intense isolation. Any difference is one of degree. The aggregate of genes may be huge, as in a whole breed, or progressively smaller as fewer and fewer animals provide the aggregate until, in the nearly extreme case, it becomes that of the pair of animals which, as parents of the favoured animal, provided its genotype.

When, for example, an animal such as Favourite (Fig. 4) becomes the favoured animal and is mated again and again with his own progeny the range for possible recombinations is extremely restricted. The breeding group is dominated by a single animal, and is thus isolated further until the principal aggregate of genes from which samples can be drawn is that carried by him. This results, broadly, in all animals of these intensely isolated groups becoming related and so similar to the highly favoured animal.

End result of the process depends wholly on the nature of the gene aggregate that is being sampled. There are innumerable possible combinations, many of which may be disadvantageous and give rise to progeny very different from the favoured animal. If these are not self-eliminating, they are culled. The greatest deficiency of historical accounts, such as those reviewed in the early sections, is that they do not record these unfortunate combinations. There are references to low fertility and to loss of vigour, but it is probable that even these do not disclose the true position.

Little, of Corriedale fame, at one time inbred a Romney flock for many generations in New Zealand. He bred back repeatedly to a ram and a ewe that he valued highly. Although the progeny of his flock included some outstanding animals, there were also many so sickly and senseless that they had to be taught even to suckle. When he introduced new rams these difficulties were overcome. In recent years many kinds of animals, particularly swine, have been inbred in the U.S.A. Most of these attempts have completely broken down in the first few generations.

Such unfortunate gene combinations are the normal and frequent result of the inbreeding always associated with intense isolation. They are not evident in the parental types because the characters depend upon recessive genes (see p. 81). These are carried by heterozygous individuals which, when mated, allow the recessive genes to combine in a paired or homozygous state and, as such, to be demonstrable.

Genes of this nature are widespread but their presence or absence in particular groups of breeding stock can only be determined by experiment. Among plants, maize is one of the few that are extremely interesting to animal breeders. While many others are self-fertilized, maize is normally cross-pollinated. Every grain on a cob has the same mother, but its male parent may be any plant in the field. In an endeavour to develop pure lines of maize, such as are found in cereals, American plant breeders self-fertilized thousands of plants. In no instance did they secure outstanding progeny. These did not occur until the inbred lines were crossed. Although seed from the crossbred plants was genetically unstable, it provided such excellent immediate crops that an enormous volume of work was undertaken, all of which sought to isolate inbred lines. These were not expected to be advantageous of themselves, but only to give maximal results when they lost their own identity by crossbreeding. The programme was designed to exploit the resultant "hybrid vigour". There have been very many thousands of inbred lines developed but not one, of itself, has been outstanding, and only about forty or fifty have given maximal results when crossbred. Nevertheless, in 1942 the United States Government estimated that the use of hybrid seed corn had increased the nation's maize crop by 300,000,000 bushels.

Jenkins (1936) summarizes corn-improvement work by hybridizing as follows:

> Crosses between promising inbred strains have given marked increases in yield varying from 10 to 30 per cent above the present varieties. Inbred strains have been developed which differ in resistance to disease, insect injury, cold, drought, composition of the plant and grain and productivity, but practically nothing has been accomplished in transferring these desirable qualities from one line to another. The high increase in yield holds only for the first hybrid generation. The best results have been obtained through two-way crosses, i.e. four inbred strains have been

combined by crossing F_1 hybrids, each resulting from the crosses of two distinct inbred lines. At present (1936) 44 different hybrids have been released for commercial production.

Lindstrom (1939) suggests that a possible cause for the inbred lines being poor types is intensity of the early inbreeding. He supported his suggestion by stating that in his own laboratory he had satisfactorily developed inbred strains of mice, guinea-pigs, and fowls, without loss of vigour or fertility. His policy was to begin with low degrees of inbreeding and proceed to higher. By doing so he reduced the fixation rate for genes within the group and retained flexibility for a longer time (see p. 31).

On the other hand Miss King has established a famous colony of inbred rats at the Wistar Institute. However, from its inception each generation was derived from animals selected for size until now they are relatively enormous. In the first six to seven generations many abnormalities appeared because recessive and undesirable characters segregated. Selection was maintained for fertility, longevity, body size and vigour. Animals found to be transmitting other than these characters were rigorously culled. Males of the 36th to 38th generation of inbreds weighed 12·3 ounces at one year, whereas those of the non-inbred colony, in which the inbred lines originated, weighed 3·8 ounces. The average weights in ounces of the inbred litters were at the 1st to the 25th generation 7·39, and at the 26th to 40th generation 7·37. Non-inbred litters at these times were 6·75 and 6·23 ounces. The inbred stocks are stated to have become relatively pure after ten generations of brother and sister matings. This represents about 45 years in cow generations and a coefficient of inbreeding of approximately 90 per cent.

Inbreeding originates nothing, but it analyses and thus discovers many things. It promotes uniformity when practised among the sub-groups to which it at first gives rise. Broadly, it produces strains of animals which, because of their relatively high degree of homozygosity, are more suited for outcrossing than for continued maintenance. However, if the homozygous state were in recessive genes as, for example, appears possible in the Orlov trotters (p. 54), they would not be prepotent nor would they influence their heterozygous progeny. Characteristically, inbred strains have small size and low vigour and very frequently, as the result of wastage, they are costly to develop. Nevertheless, there is a possibility that the method may result in a uniform strain that is perfectly normal in all respects. One such has been developed by Hammond from Large Yorkshire pigs at Cambridge. Another has been reported by McMeekan from New Zealand. He described a herd of Friesian cattle, all descended from one bull and ten cows imported twenty-five years ago. The herd is said to take a leading place for high production, and the uniformity of performance among its present-day descendants is outstanding. These and other instances occur because the original gene aggregate contained so few disadvantageous genes that characters dependent on their unavoidable and coincident fixation did not ruin the venture.

Degree of Inbreeding

Isolation has been discussed as a fundamental process that may increase in intensity. The conception of intensity is expressed by different phrases. Thus attempts are made to differentiate between line-breeding and inbreeding. If there is any justification for this, it should emphasize the qualification "line". For example, the Orlov trotters (Fig. 25) were line-bred to Barss I; whereas the bull Mountain Chief (Fig. 8), although inbred, had more general ancestry. However, if it were possible to discuss this matter with Mr Richard Booth, it is probable that he would say the bull was line-bred to Crown Prince. The choice of phraseology remains optional, but the importance of intensity is unequivocal.

Wright has evolved a method of estimating the degree of inbreeding as an approach to homozygosis. The measure has been used in the earlier sections and the method of calculation is shown in the appendix. It has reference only to genes that are simply additive in their effects. However, many characters of economic value are apparently of this nature, and consideration of other possibilities would become too complicated for this discussion.

Wright's coefficient was described by him as a path coefficient. It assesses the number of paths by which any particular gene could descend from a recurrent ancestor to its progeny, close or remote, through the sire and through the dam of the animal under consideration. A recurrent ancestor is one that appears in the sire's pedigree and also in that of the dam. Shortcomings of the method are implied in the nature of polygenic inheritance, but, despite these, it is the only reasonable method of assessing the degree of inbreeding. It translates into a convenient figure what is evident in a pedigree and as such it serves a very useful purpose. It will be shown later that the dangers of inbreeding, from a practical point of view, are associated with its degree. Broadly, the more intense the greater the risk of breakdown. Ship's officers know that it is possible to move forward cautiously in unknown seas provided the lead is kept swinging to measure the depth of water. Wright's coefficient can be used for a similar purpose in navigating, by inbreeding, to an objective in the uncharted aggregates of genes that constitute the genotypes of animals.

Uniformity

Assuming for the moment that it were possible, without loss of vigour, fertility or any other essential character, to develop an absolutely uniform flock or herd, the question arises, "Would this be advantageous?" The question is implied also by the use of artificial insemination which permits an abnormally large part of a population to be related and so similar to the donor of the sperms. The more extensively his spermatozoa are used by the method, the greater is the possibility of general similarity and so of uniformity. In this respect artificial insemination is another form of isolation. Females are isolated from males other than the donor.

It is possible that a high degree of uniformity in some characters might be advantageous. For example, if all dairy cows uniformly produced large quantities of milk or butterfat, or if all sheep produced maximal wool clips of the highest value, there would be no apparent disadvantage in these peculiarities. But each animal is a composite of many characters, and the greater the number that are essential, the more difficult their fixation in one line becomes (see p. 136). Further, most markets require a wide range of types in any commodity. The point is well illustrated by the large number within wool even when confined to that from Merinos. All of these types are absorbed annually and restriction to any one would be disconcerting to purchasers. Moreover, no individual animal or group of animals, even those now fixed in laboratories, is perfect, and over-emphasis upon one or a few characters may lose the substance for a shadow. Again, no known strain or breed of animals is ideal for all situations or conditions. Rightly or wrongly, the Australian Merino has had many "ideal" types in its history. This is true of most breeds of domestic animals, as is shown, for instance, by the earlier Shorthorn history. Further, compensatory mating, that is, the mating of unlike types, will always play an important part in any realistic animal breeding programme regardless of the breed or strain involved.

Experience shows that inbreeding processes, which are inseparable from the production of uniformity, frequently give rise to considerable economic loss when they are implemented. For all these reasons, undue emphasis upon uniformity is probably unwise. Its only possible advantage is in inbred strains designed especially for outcrossing. If the hypothetical inbred group at its highest state of perfection were capable of reproducing indefinitely without deterioration, it would still remain only an isolated unit in a huge population. To achieve any other than restricted success, it would need to be used for what would amount to crossbreeding with the remainder of the population. It would then lose its identity in the progeny of its members. This was the fate of Bates's Duchess Shorthorns and of Bakewell's Leicester sheep. Further, results from such cross-matings cannot be predicted from characters of the inbred group. Only a few will combine advantageously and another experimental period is necessary to explore the possibilities.

There is an instance of this in a colony of inbred guinea-pigs within the United States Department of Agriculture Bureau of Animal Industry. They have been mated, principally as brother to sister, without selection for a number of generations. Originally there were twenty-three families, but after twenty-five years only five remained. Members of each of the remaining families were so similar that the author, as a visitor, had no difficulty in determining the appropriate family for particular individuals. Their family characteristics were obvious. Eaton (1941) reported results from cross-mating these families. There could not be more suitable material for such an experiment. Nevertheless, he found that different sets of genes governed similar characteristics in the different lines and performance of the hybrids could not be predicted from performance of the inbred families. He stated that

> no one strain was superior in all characteristics. In crosses between strains of high and low fertility, the hybrid was usually higher in fertility than the poorer parent but lower than the better parent. The greatest improvement in fertility was obtained when two strains medium in this respect were crossed. The same relationship held for growth as for fertility. The results in general seem consistent with the manifestations of heterosis (hybrid vigour) as found in hybrid corn and in other plants and animals.

Although complete diversity is not desirable, it is evident from the foregoing that absolute uniformity, with its inseparable problems and uncertainties, would be at least equally undesirable. However, restriction of the range of dissimilarity is advantageous. This can only be maintained when inbreeding is used rationally within limits predetermined by experiment in particular breed ing groups. Hence the importance of some yardstick to measure and compare its degree.

CHAPTER XII

MODERN ANIMAL BREEDING

POLLEDNESS IN CATTLE

SEX-DETERMINATION in poultry and modification of the two new polled breeds of beef cattle are probably the most direct applications of the genetical principles reviewed. The first has been discussed (p. 81); the other is associated with the improvement of polled Herefords and Shorthorns.

These two kinds of cattle are strains of their parental breeds. Both are stated to have had their origin in polled types discovered among fully registered horned Herefords and Shorthorns. Descendants of the polled cattle were accepted for registration, and for some time it was necessary to concentrate upon polledness until the character was fixed in a sufficient number of cattle. It then became evident that, excluding genes for polledness, the remainder of the gene aggregate isolated by the polled types did not have all the advantageous genes present in horned cattle of the same breed.

This is the point at which Mendelian principles were employed. The independent inheritance of particular characters made the problem divisible into two parts, that of retaining genes for polledness and that of adding to them those for the superior carcass quality and shape that had been secured in the much larger horned section of the breed. Breeders of polled Herefords and polled Shorthorns, therefore, crossed back into the horned breeds in order to introduce genes responsible for something of a type that would be more advantageous.

They were able to do this and still retain their hold upon the genes for polledness because the condition, for all practical purposes, is inherited as a simple unit-factor character in which polledness is dominant to hornedness. The pair of characters thus conformed in all respects with the behaviour of tallness and dwarfism investigated by Mendel and discussed in Chapter VIII. A similar problem, but the converse of that faced by the American breeders, confronts Australian cattlemen who desire to breed horns off their cattle.

Polledness being the dominant condition, polled cattle can be either homozygous or heterozygous. On the other hand, because hornedness is the recessive condition, all horned cattle are homozygous. The three genotypes are *HH* for the homozygous polled, *Hh* for the heterozygous polled and *hh* for the homozygous horned beasts. In each of these formulae, *H* and *h* represent the genes for polledness and hornedness in a single or haploid state, as in gametes. When germinal cells containing them fuse, the genes will be in the double or diploid state, as in the zygote. This gives rise to the individual, and the nature of the genes in the zygote is responsible for the condition of polledness or hornedness in the individual.

When a polled bull is purchased, he may be homozygous (*HH*) or heterozygous (*Hh*). If his pedigree shows all polled ancestry it is almost certain that he will be of the *HH* type. If he is the first cross of polled and horned he will be of the *Hh* type. This may be so also if he is the progeny of a relatively recent cross of horned and polled. His breeder should be able to supply sufficient information to settle these questions, but in any case his true state will disclose itself. Presumably he was purchased to mate with horned cows; these being homozygous for the recessive condition, the mating provides the progeny-test already discussed to discover heterozygous bearers of such genes (p. 126). If the bull leaves any horned calves he will be heterozygous. If he leaves no horned calves he will be homozygous. This is a state of perfection. Frequently an animal which by all counts should be homozygous has a very few calves with butts or scurrs. Much more rarely some may have more normal horn formation.

Polled calves left by the homozygous (*HH*) bull from horned cows will be heterozygous (*Hh*) as will those by the other or heterozygous (*Hh*) bull. The horned calves left by the second (*Hh*) bull will be normal horned cattle (*hh*). However, while the first, or *HH*, bull leaves practically 100 per cent of polled progeny, the second, being *Hh*, can be expected to leave only 50 per cent of polled calves from horned cows. These percentages require sufficiently large numbers of progeny before they can become evident (p. 76).

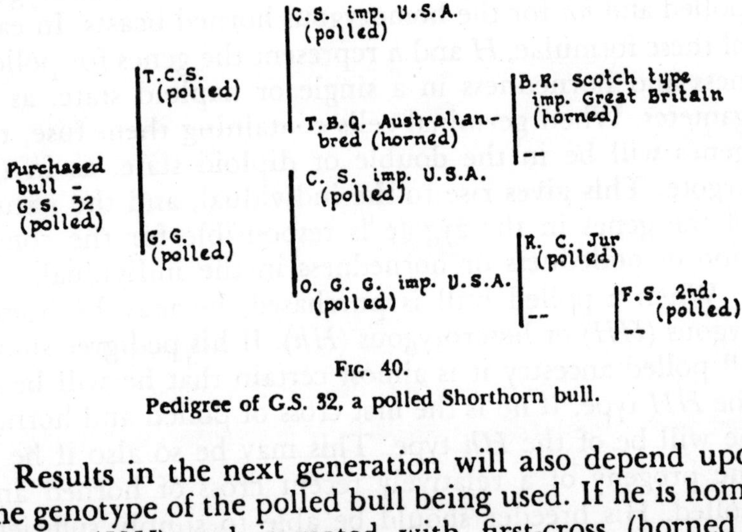

FIG. 40.
Pedigree of G.S. 32, a polled Shorthorn bull.

Results in the next generation will also depend upon the genotype of the polled bull being used. If he is homozygous (*HH*) and is mated with first-cross (horned x polled, *Hh*) cows, all the calves will be polled. If he is heterozygous (*Hh*), as he would be if he were retained from the first-cross mating, then when paired with the first-cross (*Hh*) cows he will leave approximately 25 per cent of horned progeny and disclose the typical 3:1 ratio (p. 79). When the problem is that of the American breeders, namely to introduce greater quality into polled herds by mating them with horned cattle, an opposite procedure will be followed. In each case the heterozygous animal is the crux of the position and the only way to identify or eliminate it is to progeny-test it with the double recessive, that is, to mate it with horned types and so examine the genotype of the animal in question.

The recent purchase of a polled Shorthorn bull by a leading Australian stud-master illustrates the application of these principles. The bull's pedigree, shown in Fig. 40, is reconstructed from the report of the sale which appeared in the country press.

The evidence is incomplete but probably the cow G.G. and the bull C.S. were homozygous (*HH*) for polledness. This will be assumed. The bull T.C.S., being from a horned cow (*hh*), would be heterozygous (*Hh*). The nature of his progeny would vary with the genetic constitutions of his mates. When mated with G.G., assumed to be homozygous (*HH*), all would be polled but only 50 per cent would be homozygous (*HH*). The remainder would be heterozygous (*Hh*). G.S.32, being one of these progeny, thus has a fifty-fifty chance of having a genotype which could give rise to some horned progeny. Their occurrence will depend upon the kind of mate he is given.

It was stated in the press report that G.S.32 was to be mated with progeny of R.V., an imported American polled Shorthorn bull. If R.V.'s progeny were from horned cows they would be heterozygous (*Hh*) and some of their calves by G.S.32 (*Hh*) would inevitably be horned. The possibilities could be worked out if additional pedigrees were available, but they will also be disclosed by the actual matings. The interesting point is that the American polled character is being grafted on to the Scotch Shorthorn type represented by the imported bull B.R.

DAIRY CATTLE AT DAVIS, UNIVERSITY OF CALIFORNIA

The method used at Davis combines progeny testing and inbreeding. The experiment was designed originally to investigate the genetic constitution of dairy cattle by observing results of continuous inbreeding. The process disclosed six recessive conditions: dwarfism, achondroplasia, flexed pasterns, two eye defects and lethal spasms.

During the experiment Jersey bulls were mated with Jersey cows and judged by the production levels of their daughters. Bulls that transmitted high production "rather uniformly" were then concentrated upon by inbreeding to them. They were mated with their daughters, with their granddaughters, and, if they remained alive, with their great granddaughters, as is shown in Fig. 41.

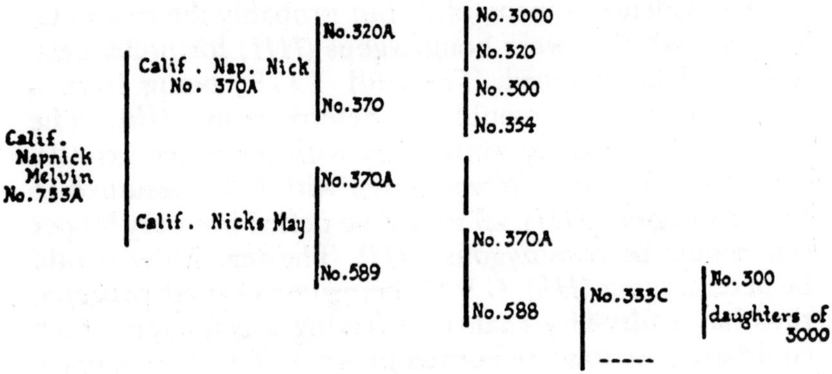

FIG. 41.

Scheme for inbreeding at Davis, California. Pedigree of California Napnick Melvin, born October 1941, herd sire, 1944.

The pedigree shows No. 753A to have three crosses of the bull 370A. This recurrent sire has had 23 daughters tested in the University herd. Their average mature equivalent was 737 lb. of butterfat. His 49 daughters tested in three co-operating herds, as two year olds, in 305 day lactations and milked twice daily, averaged 7348 lb. of milk and 423 lb. of fat from dams which averaged 6232 lb. of milk and 336 lb. of fat. The young bull No. 753A (Fig. 41) has a Wright's coefficient of inbreeding of 38·6 per cent, 37·5 per cent of which is to 370A.

Many bulls bred in the experiment have been used in commercial dairy herds co-opted for the purpose. By moving the bulls from one herd to another 29 of them have been used in 83 herds. One bull worked in seven herds and 17 in three or more. Figures are available for 29 herds and in them overall average uplift over the dams' records has been 376 lb. of milk, 0·12 per cent of fat.

and 27 lb. of fat in 305 days, all figures being derived from two-year-olds.

Daughters of 22 out of 29 bulls exceeded production of their dams. Such bulls can be expected to be about 30 per cent of all bulls selected by usual criteria for service in dairy herds (p. 103) and occurrence of 76 per cent cannot be accidental. It results from the degree of uniformity begotten by the method of breeding, and over one hundred per cent increase in bulls of that character is vitally important.

The foregoing reports success of the plan; it does not, however, disclose the whole position. From personal observation and from the description of Professor Wilson of California University, there is no doubt that, in his own words, "some of the families just pooped out"; they could not stand up to the inbreeding.

Data from the experiment are not yet published, but two very interesting points are disclosed by Professor Wilson, who very kindly forwarded some notes upon them. The first is that increased mortality up to four months of age, the criterion of vitality, was associated directly with increasing coefficients of inbreeding determined by Wright's method. Thus "coefficients up to 14·9 per cent significantly increase the percentage of mortality. Inbreeding coefficients of from 25 to 35 per cent increase it still more and inbreeding coefficients over 35 per cent give still higher percentages of mortality". It will be remembered that a previous section (p. 141) referred to the use of this measure of the degree of inbreeding as a ship's officer uses the lead. These observations confirm that suggestion and will be considered again.

The second point of interest is that the percentages for mortality were true only when data based upon the combined figures for all bulls were considered. Results from individual bulls varied. They showed "that the percentage of mortality of the inbred progeny is dependent upon the genetic constitution of the bull that is being inbred". It will be remembered *that when the data in*

Table X were being discussed a similar general effect was observable, but that differences ascribed to the genotype of particular sires were also disclosed. These two observations emphasize the importance of individual genotypes as distinct from those responsible for mass actions. No matter how thoroughly the general position may be studied, genotypes of individuals, and so particular samples of the aggregate, are the critical units.

SIB*, OR FAMILY METHOD AT MOUNT HOPE FARM, U.S.A.

The sib method arranges for a continuous series of progeny-tests within the same population of animals. It is based upon variations that take place among progeny of the same parents (p. 85). Goodale (1938) states that the Mount Hope method "seeks to identify, and use to best advantage, individuals which demonstrate their ability to promote progress in the direction toward which selection is being carried out".

This method is used principally in poultry but also in dairy cattle and more recently in mice, as an ultimate test of the method. Males and females are tested, permitting superior males to be mated with progeny of superior females. It is thus a "two way" application of progeny testing. The unit of comparison is the average of characterization within the family of two parents where "family", as in human relations, connotes progeny of two individuals.

In any year, about half the poultry matings are between cocks and hens already proven with regard to their ability to transmit the character under consideration. The other half of the breeding pens is filled with paired cockerel x pullet matings designed to test young birds. Experience has shown that only about one-third of these birds will qualify under the test, hence the large number of pens. Every hen is trap nested and a system of weighing has been implemented whereby the weights of eggs are

* All offspring of the same two parents are called sibs, or full-sibs, regardless of the sex of the progeny. They may also be referred to as being sib to one another Similarly, animals may be half-sibs, or half-sib to one another, when they have one common parent.

sampled on the 50th and 200th days. The 50th to the 60th eggs and the 200th to the 210th eggs are weighed and the average of each lot of 10 is recorded. In addition, at frequent intervals, all eggs that are laid in one day are weighed to get the flock average. Every hen is weighed when she lays her first, her 50th, and her 200th egg. There are a number of other records with regard to rate of laying, number of eggs in winter, summer and autumn periods, and duration of winter pause.

The breeding plan can be stated as if implemented in several stages. Foundation birds: A number of male and female pairs is selected by all available criteria. Each pair becomes the parents of a family and all its progeny are full-sibs. They constitute a sib-group.

Testing sib-groups: Each of the sib-groups or families, and so each pair of parents, is classified by performance. All individuals in every sib-group are tested. It is the unit to which classification refers. The criterion for classification of the sib-groups is an improvement as a whole over the performance of their parents or of families in the earlier generation.

Multiplication: Cocks and hens of proven ability are mated to reproduce maximal numbers.

Points made by originators of the method are: (i) Individuals to be parents are selected, not only on their phenotypes, but also on the basis of their full-sibs. This is done in order to take advantage of differences in parental genotypes and to increase the chances of securing superior genotypes for testing. (ii) Individuals having undesirable genotypes are separated from those with desirable genotypes by the character of their early progeny. (iii) Individuals with desirable genotypes are mated together to obtain, for testing, additional offspring that are moving in the direction desired by the selector.

Progeny testing of dairy cattle is regarded as a second application of the method suited to species in which females have relatively few progeny.

Goodale, who is geneticist at Mount Hope farm and reported the method, states that its second form is an adaptation of the method used in a classic and successful maize-breeding experiment at Illinois.* The maize-cobs are the equivalent of the families or sib-groups (p. 150). Seeds from selected cobs are individuals from groups showing the greatest movement in the desired directions. They became parents of the next generation. Because the family of a paired bull and cow is small, the sib method, in its first form, cannot be used; but the bull can be tested by the large family it is possible for him to beget, all of which are half-sibs. His progeny group is the equivalent of the corn-cob, every grain of which has a common mother but any sire. The method in its second form, as applied at Mount Hope, is to use only bulls with high indices for production (p. 122). These follow one after another in the herd.

Similarity between the Mount Hope and the Californian procedures for cattle breeding is apparent. However, at Mount Hope nothing is stated with regard to deliberate attempts at fixation by inbreeding. Evidently, progressive improvement is expected by the repeated accumulation of genes for the characters being selected. In its first form, however, there is continuous genotypic selection presumably within the same population. The methods used for the fowls and the mice are similar. Although a degree of inbreeding for these was not stated, Goodale, in 1941, reported that 24 generations of mice, involving some 25,000 individuals, had been derived from a foundation stock of 16 mice and that, therefore, they were interrelated. This suggests that the fowls also would be inbred to some extent.

Results of the method have been outstanding. Each of several characters of laying hens was concentrated upon independently but data with regard to many others were

* The Illinois experiment, which extended over twenty-three years, investigated the possibility of selecting and propagating maize high or low in protein and oil when compared with the original plants. The method was to select cobs, generation after generation, that assayed high or low and to plant grains from them. The original contents of protein and of oil, respectively, were 10·9 and 4·7 per cent. The end result was a high protein line of 15·5 per cent, a low protein line containing 7·3 per cent; a line with high oil content of 9·4 per cent and a low line of 1·9 per cent of oil.

collected. Egg number was the first character selected for improvement. Within six years the 168 egg average for production per bird was raised to 220 although management and other environmental conditions remained constant. Egg weight was isolated for improvement in 1929 when the average weight of all eggs was 23·4 oz. per dozen. Egg number was held at the high level and by 1936 egg weight had been lifted to 26·0 oz. per dozen above which figure there was considered to be no commercial advantage. Attention was then concentrated upon mortality, which is defined as death due to disease or degeneration which develops within the bird's own body without aid of an external agent of any kind.

From all causes under consideration, deaths during incubation, brooding, rearing and in the first year of laying were 60 per cent in 1936. By selective breeding, in flocks on the high levels of production and egg size, the figure was reduced to 25 per cent by 1940 and remained approximately stationary at that figure.

Workers at Mount Hope found increases in body weights and age at first egg to be associated with large-egg size. Further, percentage mortality, as defined, increased while selection for large-egg size was being practised but fell to its previous level when the selection for size was discontinued. Selection for egg size was not successful until the second generation which suggests it to be a recessive condition. Sons of large-egg mothers, when mated with small-egg birds did not beget large-egg progeny but, when sons of large-egg mothers were mated with females of the first cross (large-egg males x small-egg hens) then females of the second generation laid large eggs. After that, large-egg sires begat large-egg daughters when suitably mated.

Figures presented for periodic improvements show the method to have been successful, but whether the upper limit of improvement was reached is not known. For example, after 1929, when average egg number had been raised to 220 eggs per bird, selection for the character was discontinued. Similarly when egg size was increased to

26 oz. per dozen, selection for that character was also discontinued. A colony of mice was started with the object of carrying the method to its ultimate capacity among animals the high reproductive rate of which made the experiment possible. The mouse colony was of self-coloured animals until one appeared with 12 white hairs on its forehead. Thereafter selection was practised using the method to increase body weight of the mice and to increase size of the white patch. Body weights of two-months-old females have been increased by 15·6 per cent in 19 generations and quantity of white hair on the forehead has been increased from 12 white hairs to an almost completely white head—a bald-faced condition—by the method of genotypic selection.

TABLE XII

INCREASE IN MILK AND BUTTERFAT BY USE OF INDEXED SIRES

Figures Converted to a Mature Advanced Register Basis

Generation	*Number*	*Average mature equivalent advanced register records*	
		Milk (lb.)	*Butterfat (lb.)*
P	38	11·876	583
F_1	32	15·757	757
F_2	6	18·700	884

Uplift of dairy production has also been considerable. Increases of milk yield and butterfat shown in publications for the herd are those presented in Table XII.

UNITED STATES DEPARTMENT OF AGRICULTURE PROVEN SIRE PROJECT

The Bureau of Dairy Industry of the United States Department of Agriculture undertook a series of experiments in which bulls proven by the production levels of their daughters were used in herds of the Bureau. The plan had two forms. In one, a proven bull was used

TABLE XIII

DISTRIBUTION OF BUTTERFAT RECORDS OF FOUNDATION COWS AND DAUGHTERS OF 6 PROVED SIRES (Beltsville)

Classes: Pounds of butterfat	*250 to 299*	*300 to 349*	*350 to 399*	*400 to 449*	*450 to 499*	*500 to 549*	*550 to 599*	*600 to 649*	*650 to 699*	*700 to 749*	*750 to 799*	*800 to 849*	*850 to 899*	*Average*
Foundation cows	1	1	1	2	1	2	1	1	2	4	—	—	—	549
Daughters of first proved sire —Johan Woodcrest Lad 11th	—	—	1	2	1	—	3	0	1	2	1	—	—	581
Daughters (inbred) of second proved sire—Johan Woodcrest Lad 11th	—	—	—	1	—	2	1	3	1	—	—	—	—	579
Daughters of third proved sire —94-B (87½* per cent son of first proved sire)	—	—	—	1	—	3	4	4	1	4	1	1	1	641
Daughters (inbred) of fourth proved sire—94-B	—	—	—	—	2	2	2	—	—	1	2	—	—	591
Daughters of fifth proved sire (75* per cent son of 94-B) ..	—	—	—	—	—	2	2	3	1	—	2	—	—	629
Daughters of sixth proved sire (75* per cent son of 94-B) ..	—	—	—	1	1	1	2	3	—	—	—	—	—	566

* Percentages stated are for "Percentage of blood relation" (p. 132).

for his lifetime and was followed by his inbred sons. The other used proven bulls regardless of the relation they bore to one another.

The first form was similar to that used at Davis, California (p. 147), but the second was quite different. The problem investigated by it was whether genes for inheritance of high production were similar in all strains of dairy cattle. If this were so, then blood lines, pedigrees, and other aids to selection would be of secondary importance to the progeny-test. The experiments were very carefully carried out. Every female was expected to pro-

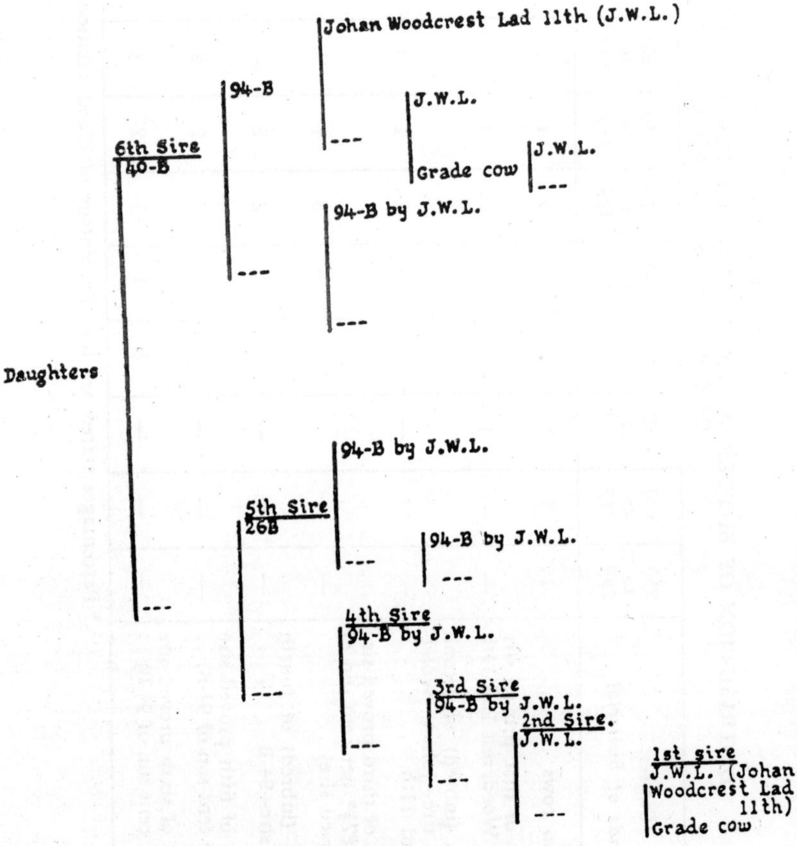

FIG. 42.

Reconstructed pedigree to show probable method of mating back to Johan Woodcrest Lad 11th at Beltsville.

vide the best record of which she was capable, and in certain of the herds all cows were kept long enough to produce a heifer calf. Conditions of production were standardized with regard to management and food, so that records would be comparable throughout.

First Form

Graves and others (1936) reported upon work carried out at Beltsville with a Friesian bull, Johan Woodcrest Lad 11th, and his progeny. They state that "fortunately he carried no factors for abnormal characters that would have been intensified in his inbred progeny". This bull was mated with grade Jersey, Guernsey, and Holstein cows and was bred back on his own daughters from them. He was followed by his son, 94-B, out of a cow with 75 percentage-of-blood relation (p. 132) to Johan Woodcrest Lad 11th.

Results from this form of the experiment are reproduced from the original report in Table XIII. It shows butterfat records of foundation cows and of daughters of the six sires used.

The pedigree (Fig. 42) is reconstructed from information given in Table XIII, and from that which follows.

The foundation cows used in the experiment had an average butterfat yield of 549 lb. The percentage of foundation cows, and of daughters of the various sires that had records of 550 lb. of butterfat or above, are as follows:

	Per cent
Foundation cows	50
Daughters of sire No. 1, outbred	63
Daughters of sire No. 1, inbred	68
Daughters of sire 94-B, outbred	80
Daughters of sire 94-B, inbred	56
Daughters of sire 26-B	80
Daughters of sire 40-B	62

The terms "outbred" and "inbred" differentiate between first-cross (50 per cent) daughters and daughters bred back to the sire (75 per cent). Some designated as outbred daughters of 94-B were closely bred, being out of the outbred and inbred daughters of his sire, who was also, on the maternal side, his grandsire and great grandsire. The tabulation shows a consistent elimination of lower level records. Exceptions are daughters that resulted from mating 94-B's own daughters back to him, and the daughters of 40-B.

The report of the work by Graves states:

> On the whole the results of inbreeding in this experiment have been quite successful, owing largely to the fact that the first sire used, Johan Woodcrest Lad 11th, whose hereditary traits were concentrated in the succeeding generations, possessed excellent germ-plasm for producing capacity and was free of factors for lethal traits, lack of fertility, and other abnormal characters. A similar experiment, concentrating the inheritance of a sire possessing inferior germ-plasm, would have had very different and less favourable results.

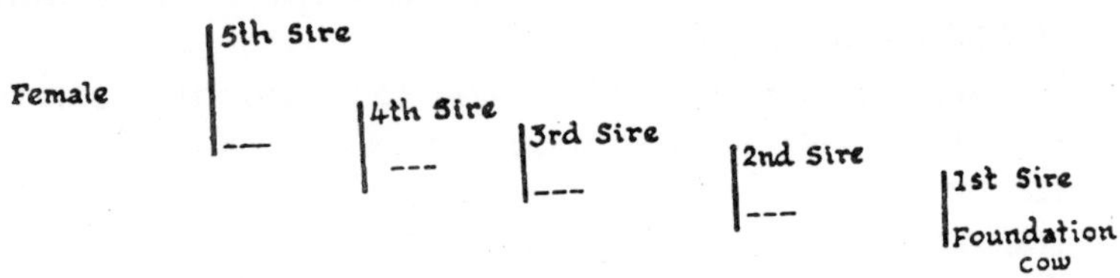

FIG. 43.
Pedigree of a sixth generation cow (second form of B.D.I. method).

Second Form

Records of cows used in the second form were also uninfluenced by selection because all females were tested. The first bull used was previously untried, but in the experimental herd at Huntley Station he proved himself capable of begetting high producers. After him each succeeding bull was proven, often in outside herds, before being used. Five unrelated bulls comprised a series of proven sires each following another.

Results show that the general trend of production within the herd has been upward. This, Graves states, "is partly shown by the percentage of records of 600 lb. of butterfat or better for the various groups". These percentages are:

	Per cent
Foundation cows	46·3
Daughters of 1st sire	77·7
Daughters of 2nd sire	64·3
Daughters of 3rd sire	54·5
Daughters of 4th sire	75·0
Daughters of 5th sire	80·0

A reconstructed pedigree of a cow in the sixth generation of the herd is shown in Fig. 43.

During the experiments, which began between 1918 and 1920 and were reported upon as above in 1936, a number of probabilities have been isolated as tentative conclusions. Categorically they are:

1. From the experimental work, and other evidence, it appears that meritorious sires within a breed have the same genetic make-up.

2. When such sires are adequately and definitely proved, mating with the daughters of each other will be followed by good results although the sires are unrelated.

3. Selection for herd duty of young bulls that are sons of bulls definitely proven to be meritorious and have dams and other female ancestors all by meritorious sires would ensure, with some certainty, that the untried bulls had germ-plasm likely to give rise to high production.

4. The number of successive sires of proven merit, in sequence uninterrupted by unproven or poor bulls, is most important in building up superior inheritance in dairy cows.

5. In herds that have used twelve proven sires, there may be no female line with more than four successive crosses of these. Such a herd has made no greater progress in accumulating superior inheritance than one which has had only four proved sires but has used them so that all appear consecutively in several female lines of descent.

6. Progress made in concentrating superior inheritance may be estimated by comparing the number of successive crosses of proven sires in the female lines with the total number of sires of proven merit used in the herd.

Bulls born in the Bureau of Dairy Industry Huntley Station herd have been placed with cows of farmer co-operators to prove out their transmitting ability for levels of production. There are fifteen of these co-operating herds that have used three or more station-bred bulls where each bull has three or more daughters that have completed a lactation period. These bulls are the product of one to five generations of proved sire matings and are put into use in the co-operator's herd when old enough for service.

When data of the fifteen herds were brought together and summarized, it was learnt that there were 265 foundation cows whose best records averaged 411 lb. of butterfat; 265 first generation daughters carrying one cross of station-bred bulls, whose best records averaged 461 lb. of butterfat; 171 second generation daughters, carrying two crosses of station-bred bulls, whose best records average 486 lb.; 67 third generation daughters, carrying three crosses of station-bred bulls, whose best records average 541 lb. and there were 26 fourth generation daughters, carrying four crosses of station-bred bulls, whose best records average 559 lb. of butterfat.

Results of this work should be associated with those of the Davis experiment (p. 147). The Bureau of Dairying project shows how ceiling levels of production can be uplifted by using generation after generation of proven sires. The Davis experiment demonstrates that genes for high production can be fixed, in the genetical sense, until the percentage of bulls capable of uplifting production levels of their daughters is increased over one-hundred per cent. These two findings should be of vital importance to the dairying industry.

THE U.S.A. BUREAU OF ANIMAL INDUSTRY PROJECTS IN BEEF CATTLE, PIGS AND SHEEP

As indicative of modern thought with regard to animal breeding, activities of the Bureau of Animal Industry*

---

* The United States Department of Agriculture has two Bureaux which investigate animal problems. One is restricted to those of the dairy industry whilst the other deals with all the remaining animal industries. and has a Division of Animal Husbandry.

show two major trends. One is to find methods of exact measurement for characters of economic importance. The other is to undertake improvement of breeds and strains of animals by processes used with success in the Bureau of Plant Industry for maize, which, under natural conditions, is a similar crossbreeding organism.

Exact Measurement

Mathematical processes evolved by Fisher and others were stated to have opened new possibilities for advancement. Broadly, they differentiate between the effects of a complex series of causes. They do this by the analysis of variance, that is, of observed differences in aggregate effects. However, such differences must be expressed by figures before the processes can be used. By their use it is possible to detect and to isolate individuals and strains which carry inherent factors for high production. They are able to do this because they can attribute to hereditary, environmental, and other causes their proportionate effects.

When this is done and superior animals are discovered, there is ample genetical knowledge available to make it possible to establish, eliminate, or multiply particular individuals, but measurement is an essential preliminary to investigation. Genetical improvement is not accidentally confined to milk-yield and butterfat or to egg number and weight. It has been made possible because these characters are readily measurable and the genetical significance of variations can be made clear.

The foregoing statements to a large extent paraphrase those made by Dr H. C. McPhee, Chief of the Division of Animal Husbandry, Bureau of Animal Industry, in 1936 and in 1939 when he was discussing the work of his Division. It is not surprising, therefore, to find that for some years many of its workers have been attempting to evolve methods of measuring factors that make for quality and palatability of meat, for characters of wool, for vitality, vigour, and other similar units of economic importance.

Animal Breeding

The most recent development is to organize a number of regional breeding laboratories. These have been opened for beef cattle, pigs, sheep, and poultry.

The main objective for swine is improvement by the application of breeding methods. Improvement is being sought in production of sows, in growth of pigs, in economy of gains, in physical vigour, and in quality of carcass. To accomplish these objectives, use is being made of production recording and progeny-testing in combination with various inbreeding and outbreeding systems.

Thirty-six separate strains are being developed by close breeding. Twenty-three in the Poland-China breed, ten in the Duroc Jersey, two in the Hampshire, and one from a Landrace x Tamworth crossbred foundation. Eventually these inbred lines will be crossed. A Landrace x Black Hampshire cross has been made on a scale which will furnish needed genetic information on various characters in addition to establishing the basis of a new strain. Similar crossbreeding experiments are being carried on with Landrace x Poland-China, Landrace x Duroc Jersey, Landrace x English Large Black, and Yorkshire x Duroc Jersey. Landrace pigs for these purposes were imported from Denmark. All of these crosses are carried either to the F_2 or to back-cross generations and in numbers approximating 150 for each cross. In addition, one inbred strain is being developed in each of the Danish Landrace, Duroc Jersey and Poland-China breeds. In all, forty-four separate strains of swine are being developed. Thirty-eight of these are confined to the so-called pure breeds and six have a hybrid foundation.

The objective in sheep is to bring about breeding improvement for wool- and lamb-producing qualities, using about the same scheme of close breeding as for swine. Brother and sister mating will be replaced by parent x offspring, full brother x sister, half-brother x sister, and other types of mating. The Rambouillet is the starting point. The first matings were made in the autumn of 1937. A number of families of twenty-five ewes and one ram were divided approximately equally for allotment to three projects: close inbreeding, moderate inbreeding, and the proving of sires. In 1938 the programme was expanded to a total of sixty-four separate breeding pens. It is planned in the future to test sires by means of an inbreeding programme before new lines are started.

The Bureau accepts that outstanding breeding lines and proved strains can be developed by inbreeding. It proposes, when they are evolved, to combine them to take advantage of hybrid vigour. It is probable that attempts will be made to produce three-way and double-hybrid in addition to single-hybrid lines. Simultaneously an outcross scheme will be employed to test the value of different strains in improving random-selected stock.

Extensive data are being collected on characteristics of fibres and on conformational characters of the sheep. The primary effort is to develop closely bred strains which can be crossed later and used to develop superior flocks. The problem is one of creative genetics, involving a drastic reshuffling of the genes by means of inbreeding. It is hoped that, through recombinations and hybrid vigour, a significant improvement will result in the sheep. When the above report was made in 1939, it was stated: "At the present time we do not know whether this means relatively homozygous lines or a programme of controlled heterosis."

The mechanics of these experiments were recently observed. Beef cattle breeding is being investigated at Miles City, Montana, on a station with 90 square miles of typical cattle country capable of carrying about 20 head to the square mile. However, there are "bottom lands" with heavier carrying capacity.

Record of performance studies have shown differences in weaning weights of calves, daily gain while in "feed lots", and efficiency of food utilization (i.e. gain per 100 lb. of digestible nutriment consumed) that are all relatively highly heritable characters. The proportion of lean meat to bone on the 9th, 10th and 11th rib cut is also highly heritable. These characters, all of which are economically important, are the criteria for selection within the experimental herds.

There are three "families" or "lines" of Hereford females, each with its own bull. In addition there are a number of "test cows". Bull calves born in the "families" are examined with regard to weaning weights which, with those taken at their births, give the rates of gain and so milking ability of the dams. The calves are then put on "full feed", and the results disclose rate of growth due to the calves' own ability to develop, economy of gain and conformation. All these are phenotypic criteria but, when the calves are old enough to beget progeny, they are progeny tested, i.e. selected genotypically.

All those selected for test are mated with 25 or thereabouts of the test cows. Eight steer calves are taken from each test group. They are fed out until 900 lb. live weight. Growth rate, economy of gain and body type are determined and, when the calves are slaughtered, percentage of carcass to live weight, grading of carcass and block tests are carried out. Relative merit of the young bulls is determined from the scores of their progeny with respect to utilitarian characters. None other is used.

The best of the young bulls are used in the families or "lines" in which they were bred. In these their progeny are again appraised so that the bulls have two genotypic appraisals, the second of which is made with higher quality cows.

Similar methods are used with sheep, but for them fleece characters are taken into consideration. Samples of wool are collected to determine yield, and from it and the fleece-weight, production of clean scoured wool is determined for every sheep. Other samples are taken for fibre diameter and uniformity, measurements being made with microscopic aids. Staple length is also measured. Conformation is scored and no ram is used that does not have a high score in these aspects. Those with high individual scores are test mated, as were the young bulls, and appraised by the excellence of their progeny in respect to the criteria stated above.

These projects have been under way for about three or four generations. Considerable differences have been observed between the lines with respect to one or other of the characters under observation. Eventually, after some degree of inbreeding to fix the observed characters, lines will be crossed, one with the other, as was described earlier.

PROGENY-TESTING ASSOCIATED WITH SELECTIVE REGISTRATION AND ARTIFICIAL INSEMINATION

Pig-testing in Denmark

Originally the pig industry of Denmark was established to produce pork but, when fear of disease caused England to shut her doors, the Danes were forced to look for an alternative outlet. They found this in bacon, but they first had to change the type of their pigs. An important factor in this process and in securing optimal results for producers was the Danish pig-testing system.

Testing stations were set up throughout the country. It was obligatory for each breeding centre to cause one litter for every two sows in the centre to be tested each year. Four pigs were sent from each litter to the testing station. They were fed there until they reached market weight. All environmental conditions were standardized and data were secured with regard to economy of gain, rapidity of growth, and carcass quality. Not all the pigs under test were killed. Two from each lot of four were sent back to become breeders. In this way members of superior litters were enabled to reproduce.

The progeny-tests were publicized and became the basis of evaluation. Results have been very satisfactory. In seventeen years, the average time taken for pigs to reach the 200-lb. killing-weight has been reduced by seventeen days. Also consumption of food has been reduced from 4·5 lb. to 3·3 lb. of food per pound gain in body weight. The rate of change affected by the method is shown by the average results for 1929-30 and for 1933-4:

	1929-30	*1933-4*
Food units per lb. of live weight	3·39	3·31
Weight of bacon as percentage of live weight	59·6	60·2
Thickness of back fat (inches)	1·53	1·40
Thickness of belly (inches)	1·21	1·41
Length of body	35·4	35·8
Percentage of pigs in Quality Class I ..	52	75
II ..	26	18
III ..	22	7

Selective Registration

All Advanced Registers of Merit are instances of selective registration. The method of improvement would not be regarded as modern if it were not that many breeds and some countries are only now adopting it. Nevertheless it is a practice of long standing. For example, in 1883, the Kiama (New South Wales) Agricultural Association's Herdbook was established. One of the rules for registration was that a cow must produce at least 12 lb. of butter or 350 lb. of milk in one week, or be the progeny, male or female, of cows qualified for a place in the herdbook. For a bull to be eligible, four of his progeny must have passed the required test. Another of the rules required that tests for registration should take place in the same locality, "the object being to place competing animals on an equal footing".

Denmark was a country capable of outstanding co-operative effort, and there and in Holland conditions for registration required support by evidence with regard to the merits of progeny. This is the essential difference between old and new criteria for selective registration. The first had reference to the individual's own record and thus was phenotypic. The more recent presents evidence as to the genotype of the animal. For example, here is the information available in Holland with reference to a bull, Adema's Athleet, born 24 February 1929:

The bull scored 85 points in the Friesian herdbook and 85·5 points in the Netherland herdbook. He was declared "preferent" after an examination of his progeny in 1934. When five years old he had eight daughters averaging 6962 lb. milk with 3·87 per cent fat in their first lactation. The average production of his first 100 daughters, in their first lactations, was 7586 lb. of milk with 4·05 per cent of fat in 330 days.

In Denmark presentation of superior dairy stock in show rings and similar exhibitions is accompanied by their production data. Watson (1939) who attended the Copenhagen Exhibition of 1938 stated the rules set out that no dairy cow could be shown unless she had been officially recorded for at least one full lactation of 304 days. Each cow's record was evaluated on a system of points. This started at zero for a calculated annual yield of 405 lb. of butter and one point was given for each additional 11·25 lb. Thus a cow giving an average of 585 lb. of butter would receive 16·4 points. Allowances were made for immature females. Additional points were given for the butterfat percentage of the milk (as distinct from the total yield of butter) and still others in cases where the cow had four or more lactation records.

A typical catalogue entry gave the animal's name and herdbook number, and the names and numbers of the sire, dam, grandsires and grandams. Then came a series of figures such as: "Mean 3 years (1934-37) 6323 4·65 331(363) Y 39·3 points." These figures meant that the average yield for the three years had been 6323 kilos (14,000 lb.) of milk at 4·65 per cent butterfat, that the calculated average yield of butter was 331 kilos (704 lb.) and the highest yield in any lactation 363 kilos (806 lb.). These figures gave 39·3 points for production. The highest score in the old cow class was 64·1 points for a yield of 1430 gallons at 5·61 per cent butterfat.

Apart from all this, the cows were judged as they stood for conformation. The main points for which the judges looked were size of frame, depth and width of body, and other points of "constitution". Little attention seemed to be given to minor points of conformation, and a good many high awards were given to cattle with obvious faults such as high tail-heads, slack loins, and coarse horns. The classes were not strictly competitive. No money prizes were offered, the awards consisting of gold, silver or bronze medals of which the judges appeared to be able to award as many as they chose. In large classes as many as three or four medals of each grade were awarded.

There was a similar system of double awards in the bull classes, one set of medals being given for conformation and another for pedigree. The judging on pedigree was again based on a system of points; one number, out of a maximum of 24, was awarded for the butter-yields of the dam and two grandams, and a further number, again out of a maximum of 24, for the conformation of the ancestors. The printed herdbook of the breed gives for each beast a score of points for conformation, and it was on these figures that the bull's score for "conformation pedigree" was calculated. For example, the particulars of a famous old bull were set out as follows:

Dam's average yield	5 years	16,180 lb.	at 4·15	per cent
Dam's dam's average yield	10 „	12,190	„ 4·05	„ „
Sire's dam's average yield	4 „	11,100	„ 4·06	„ „

This series of records was sufficient to earn the maximum of 24 points for pedigree of performance; the points for "conformation pedigree" were 23, so that the bull's total pedigree was valued at 47 out of 48 points.

Apart from the classes for individuals, there were groups of cows and groups of heifers from particular herds, and also groups of cows and groups of heifers by particular sires. In the herd classes the number of animals to be exhibited depended on the size of the herd. The minimal number was three, but herds containing more than forty cows had to show four and those with over eighty cows, five.

The selection of bulls is based primarily on the milk records of their daughters, and the sons of proven sires are given, in turn, an opportunity to prove themselves. The record of Hojager, for example, when thirty of his daughters had given yields that could be compared with those of their dams, was as follows:

	Milk (gallons)	*Butterfat (per cent)*	*Calculated butter-yield (lb.)*
Average of 30 dams	1053	4·26	508
Average of 30 daughters ..	1149	4·36	561

The bull's daughters had thus yielded on the average 96 gallons of milk and 53 lb. of butter more than their dams. It is by such performances and not by looks that the Danish breeder values a bull.

In Germany similar genotypic registration is practised by compulsion, and the method was being applied in occupied countries to increase output of animal products. Broadly, the method is to restrict registration to animals of approved type and production which also have progeny of similar merit.

Artificial Insemination

In Soviet Russia artificial insemination was used very extensively for what amounts to "grading-up". In parts of Europe, in the United States, and more recently in England and New Zealand, however, it is being used to permit proven sires to leave a larger number of progeny. Generally, and particularly in the United States, farmers form co-operative groups which retain veterinarians to make the inseminations. Bulls are maintained by one or more of the co-operators. This process is discussed in Chapter XIV.

Nucleus Scheme

Hagedoorn (1939) has put forward what he calls a "nucleus scheme". Being from the Low Countries, it has special reference to dairy cattle, but it could be applied to other animals. Indeed, the segregation of our Merinos into stud flocks closed to outside blood provides nucleus groups for that breed. Hagedoorn's plan suggests that a number of breeding males should be judged according to the quality of their progeny, care being taken that a random sample of their progeny is examined. Males heterozygous for lethal genes or carrying those that are disadvantageous (p. 126) are to be discarded. Only a few males are to be chosen and they are to be exclusively those that have produced highly valued offspring. Qualities of real economic worth are to be given precedence over minor points or those of speculative value.

This nucleus group, containing the highly selected males, should provide all the male breeding stock for the breed. These would be confined to the offspring of the chosen males. Briefly, it offers breeders of males that satisfy these conditions a monopoly for their production. In return breeders must accept progeny-testing as the method of selecting senior sires in the nucleus groups. Hagedoorn considers that the nucleus scheme would concentrate the effort for selection into the hands of a very few specialists who would be rewarded by their monopoly. The number in the nucleus group is not limited; any owner of suitable stock can enter it, and an owner who loses his stock of that kind must leave it.

CONTROLLED HETEROSIS

Cautious Shorthorn breeders about 1830 remained aloof from the Bates-Booth controversy, but had "a bit of both". The method of breeding about to be discussed goes one better. With regard to all other methods, it has "a little of everything, please".

Every system of breeding must have an objective. This may be larger quantities of milk or butterfat, particular kinds of wool, or grades of meat. The nature of the objective is immaterial to the scheme, provided it has been determined with sufficient realism.

Implied in the conception of an objective is that of an ideal. Since the objective and the ideal are fundamentally associated, the breeding plan must have regard to establishing and maintaining the ideal. The objective and the ideal will determine the kind of foundation animals the breeder secures. These animals should be regarded as populations or aggregate of genes, and their assessment should be based upon evidence available to appraise their genotypic worth. Methods for this have been discussed.

It may be impossible to secure foundation animals that conform to the breeder's ideal. Compensatory matings will then be necessary, as they were for Charles Colling. Such matings between unlike animals are the only means whereby genes can be introduced to produce new characters. Variability will be increased and further selection made possible when the progeny of the unlike animals are themselves mated. Emphasis should be placed upon as few characters as possible. This may involve establishing within the flock or herd two or more "families", each of which has its own relatively few strong points.

This arrangement of foundation animals will disclose that, although all are similar, in many respects the subgroups are slightly different. This degree of dissimilarity is advantageous and the breeding plan that has been called controlled heterosis is designed to maintain and control it.

Normal breeding procedure will eventually discover which of the foundation animals bear or beget outstanding progeny. These are the highly favoured animals of the scheme which centres upon their use. If such animals do not eventuate they must be purchased, because without them real progress is impossible. It is emphasized, however, that they are to be judged, not by their own performance, but by that of their progeny. They are to be selected genotypically.

These being discovered, the plan of breeding begins to take shape. It has two phases. The first has reference to a very small portion of the breeding group. The second is applicable to the flock as a whole.

The first step to be taken is to mate the highly favoured animal with his or her descendants or close relatives. The move has several objectives. It makes a further test to discover whether the favoured animal has any recessive defects (p. 126). If these occur the favoured animal loses its rating and the search begins again. If they do not occur, the degree of inbreeding then practised establishes what can be regarded as a high-water mark for inbreeding within the particular group. This has been discussed several times in earlier sections to introduce the conception of safety below known levels of inbreeding.

The test should be made with animals characteristic of the group in which subsequent breeding will take place. It strongly suggests that the highly favoured animal, having been chosen by its ability to transmit advantageous characters when normally mated, is also unlikely to do any harm if concentrated upon below the determined level. This should be defined mathematically. It can be calculated as percentage of blood relationship (p. 132) or as a coefficient of inbreeding as evolved by Wright (p. 140). These figures translate and express what is implied in the pedigree of the test mating.

The step has another advantage. The disadvantage of breeding by genotypic selection is that the highly favoured animal frequently dies before it can be used with maximal advantage. When the exploratory mating is successful, therefore, other and still more intense matings are made within this small group to establish a "reservoir" from the genotype of the favoured animal. The idea is to secure, within a small group of highly selected individuals, as high a concentration as is possible to the favoured animal, so that the reservoir can be drawn upon when the animal dies.

The second part of the scheme has more general application. It may affect the whole flock or herd, when numbers are few, or it may be applied only to one section, when the original animals were divided into families. In either case matings are arranged so that relationship to the favoured animal is increased. Whereas the first part of the plan sought to establish a high concentration in one spot, this seeks to secure fairly wide dispersion of the favoured animal's genotype.

It promotes similarity but maintains differences because the degree of concentration is low. This is essential to the whole scheme. The level of inbreeding or concentration must be always low enough to permit selection and to retain maximal vigour. Artificial insemination may be used to distribute the genotype of the favoured animal but, if used too extensively, it may defeat its own end by establishing close relationship too generally.

The objective is also to accumulate, by all appropriate methods, samples of the gene aggregate which produced the highly favoured animal. Samples of it are his own and those of all his close relatives particularly sibs and half-sibs. Accordingly, progeny of these animals are also secured and examined. Broadly, the desired relatively low level of inbreeding is maintained by using not one animal exclusively but a number of animals closely related to it. Degrees of relationship and inbreeding are not to be allowed to exceed levels previously determined to be innocuous; nor must uniformity of the flock or herd approach the point where selection is no longer possible. In each breeding group the process gives rise to animals related and so similar to the favoured animal, but not extremely so.

Periodically a half-outcross of new blood is introduced, as at Ballindalloch (p. 44). In large flocks it can be arranged between groups within the flock. This frequently happens in Merino studs closed to outside blood. Introduction of new blood as opportunities offer from time to time, and always long before some such move becomes obviously necessary to recover vitality, vigour, or fertility, is an essential feature of the scheme. It supplies a constant trickle, but not a steady stream, of new genes, and thus ensures variation.

Restriction of homozygosity by the deliberately maintained low degree of inbreeding maintains the general flock in a heterozygous state. However, heterozygosis, which is the enemy of uniformity, can in turn be restricted by periodically using members of the small but highly concentrated reservoir built up for this purpose. It is the aggregate of genes which is fundamental to the whole plan. These highly inbred animals make a return to the original line possible at any time.

Heterosis is the phenomenon of increased vigour associated with crossbreeding. It is greatest when the parents

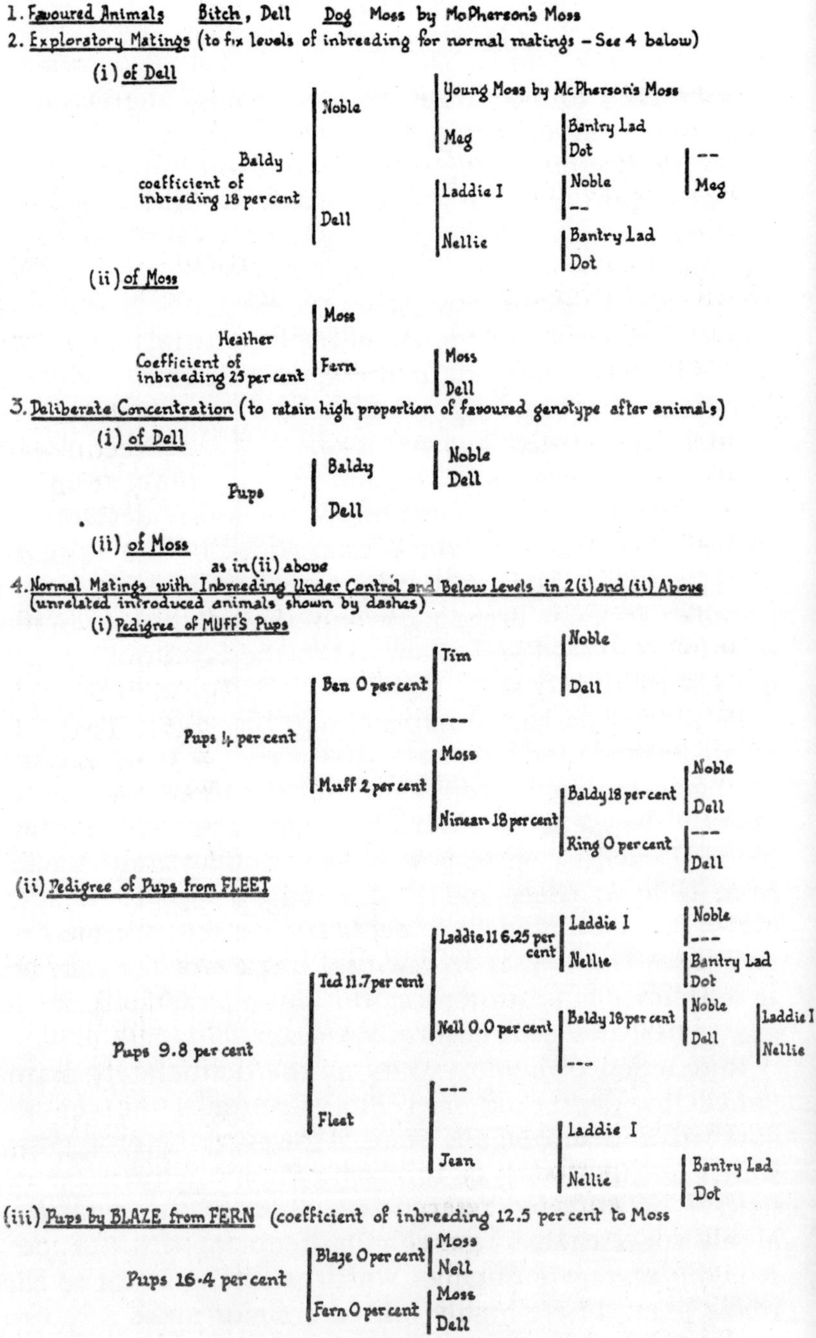

FIG. 44.

Explanatory pedigrees in method of controlled heterosis. Percentages are coefficients of inbreeding.

are most genetically diverse, and least when opposite conditions control matings. It is controlled and maintained by the see-saw methods of the plan which compel similarity while retaining slight diversity.

REVIEW OF METHODS

All the foregoing methods of animal breeding have been formulated by those fully qualified to do so by genetical knowledge.

Independent treatment of polledness and hornedness is applicable to any other pair of characters controlled by a few gene pairs. It is possible to move these at will, like pieces on a chessboard. For example, within the scheme for controlled heterosis, and from the same ancestry, two strains of Border Collie dogs have been evolved. One is prick-eared and short-coated; the other is drop-eared and long-coated. These are characters like hornedness and polledness. Similarly, red coloration, which is a simple recessive to black, has been eliminated. However, polygenic characters such as temperament, style, and method of working are controlled and uniformity produced or intermediate forms secured by the methods described for modifying and controlling groups of genes. The essential difference in the nature of these two processes, one applicable to small groups of genes and the other to large, is discussed in earlier sections.

Other methods considered were also designed to control polygenic characterization. Where figures for improvement have been available for presentation, all have disclosed success. Breakdown of the processes has been associated with attempts at fixation by inbreeding. It has not taken place in the advantageous group of genes but in others which, together with the advantageous group, constitute the genotype of the animal. For this reason the Bureau of Animal Industry now proposes to analyse and purify its inbred lines of sheep before attempting to develop them for specific purposes.

Work with proven sires, each unrelated to the other,

by indicating that large gene groups responsible for dairy production are probably as discrete as those much smaller, conforms with expectation. If this indication be confirmed and applied to other characters of similar nature, entirely new fields of work will be opened up. Just as the pair of genes responsible for hornedness and polledness can be, as it were, lifted out and put back where they are wanted, so it may be possible to lift larger groups. This is the synthetic task undertaken by the Bureau of Animal Industry in its Regional Breeding Laboratories in the United States.

The methods discussed all emphasize progeny-testing. There can be no doubt that future progress in animal breeding will result from the identification of favourable genotypes by this or allied methods. Identification will be followed by concentration upon animals which have them. Qualitative phenotypic selections will be replaced by accurate measurement of characters. Performance of animals will be the criterion of individual excellence and performances of their progeny will determine the value of breeders. The principles involved in these statements are sufficiently well established. No useful purpose would be served by waiting until every i is dotted and every t crossed with regard to the detailed mechanism of the genes responsible for characters of economic worth. Modern methods disclose obvious finger-posts to rational progress. Use of these or of others of a similar nature is fully justified. In the forthcoming struggle of wool and artificial fibres, importance attached to the progeny-testing of fine-wool sheep is so great that the process has been isolated for discussion in Chapter XV.

CHAPTER XIII

ANIMAL BREEDING IN AUSTRALIA

ENVIRONMENT AND HEREDITY

"HALF the breeding goes down the neck" is an old adage that is true when "breeding" means the final appearance of animals. For example, both Verges (1939) and McMeekan (1940) have shown, the first with lambs and the second with pigs, that the final shape and age of these animals, when they reach market-weights, can be varied greatly by modifications of the food available to them during their development.

Hammond has made very extensive studies of the relation between skeleton and form in animals raised on high or low planes of nutriment. He has shown that in the young of improved breeds, such as beef cattle or mutton sheep, the skeleton predominates—"they are all head and legs". In this respect they are similar to adults of unimproved breeds, for example, cattle of 150 years or more ago, and to wild sheep.

The early relation of bone to carcass persists if the young animals are grown with insufficient nutriment, so that they become and remain in "store condition". In such cases, essential organs grow at the expense of fat and muscle tissue, and the head, horns, and limbs form a large proportion of the carcass. However, if the animals are reared on a high plane of nutriment, the shape of the carcass and its essential proportions are very different. They are then characteristic of improved breeds.

Hammond's conception of the final state of animals and its association with the level of nutriment under which they have been reared is very closely akin to that of Goodrich (p. 69). Hammond has stated repeatedly that levels of nutriment have been critical in improving the carcass quality of animals. Thus, with reference to this conception, he stated (1942):

> Having found the optimum conditions he [the breeder] arranges to breed and select his animals under these conditions, so that he is then able to pick out and breed from those animals which show, in the highest degree, the qualities he is striving after. By this means he is able to direct the evolution of his breeds. In any but optimal conditions for production, the production is limited by the environment or food supply and not by the genetic capabilities of the animal; hence proper genetic selection cannot be made. Thus it will be seen that the breeding methods are not wholly static, in reproducing the best that has gone before, but evolutionary, in raising the level of production and quality to higher levels than ever existed previously.

Although this close association exists between levels of available nutriment and the final shape and character of animals, food is only one of many environmental forces. There are, in addition, all those collectively described as "the climate". Possibly atmospheric temperature is the most important of these. It has been shown by Rhoad (1942) and by Manresa and others (1939) that high atmospheric temperatures have specific effects upon the constituents of the blood stream of British breeds of cattle in tropical countries. The end result of these is that the cattle are unable to maintain normal levels of production.

This is particularly true for dairy production. In Wisconsin, U.S.A., winter temperatures fall far below zero and the question arose whether artificial heating of barns in which dairy cows were housed would be profitable in the winter. Arrangements were made to raise the temperature and to record production. The finding for Guernseys and Holsteins was that from 50 to 55°F. were optimal temperatures. The researchers stated (Kelley and Rupel, 1937) that such temperatures were "conducive to good appetite, body vigour and a healthful stable condition". Regan and Richardson (1938) showed that high atmospheric temperatures caused decrease in both fatty and non-fatty solids of milk and Regan, who worked under extremely well controlled conditions, was able to state 80°F. to be the critical temperature for these two and also for milk flow. It marked a decrease which rose with the temperature. Regan's figures, for pounds of milk daily at temperatures with 10° intervals from 40 to 90°F. were 29, 28, 27, 27, 25 and 20 respectively, whilst at 95°F. production fell to 17 lb. of milk per day.

Adaptation of particular breeds and strains to specific environments is well exemplified by the multiplicity of breeds in the British Isles (p. 4). Even within the relatively narrow range of environments possible in that small country, to quote Hammond (1942):

> There are breeds for beef production—the Beef Shorthorn, Hereford, Aberdeen-Angus, etc., for good lowland environmental conditions; and others—the Galloway, Highland, and Welsh—for cold, rough mountainous conditions; breeds for dual-purpose (milk-beef), the Dairy Shorthorn for good land and the Red Poll for poor, light, dry land; breeds for milk, the Ayrshire and British Friesian; and breeds for cream and butter, the Jersey and Guernsey. With sheep, the breeds range from the Welsh and Blackfaced, improved mutton breeds suited for wet and cold mountainous districts, to the Hampshire and Suffolk, improved fat-lamb breeds suited to arable areas for folding purposes.

In addition to such environmental adaptations, there are those related to causes previously classified broadly as secondary environmental forces. They have reference to varying market requirements that need corresponding modifications in animal types. Forces of this nature have been critical to all improvements in breeds of livestock from before the time of Bakewell until the present day. They were of the utmost importance in establishing Australia's wool industry. Much has rightly been said of the good work done by Macarthur and others as founders of that industry, but Roberts (1935) in his *Squatting Age in Australia* makes it very clear that economic conditions in Europe were the big factor.

After the Napoleonic wars England's export wool trade was, to use Roberts's own words, "cut away at its roots and English wool could not even hold its own in the home-markets". By 1824 English home-grown wool became unsaleable in face of competition by cheap German wools grown principally in Silesia. Duties were unable to protect the local industry. Its bedrock price for survival was 1s. 6d. per lb., whereas superior and finer German wools were available at 1s. 1d. English prices by 1825 had reached 1s. and less and never recovered. Manufacturers also found the finer German wools better suited to new trends in consumption. They said they could compete up to 1814 with British type wools but to manufacture cloths with them after 1825 meant ruin. The bottom went completely out of English wool values when 6d. per lb. was offered for its finest kinds.

When English wools seemed to have succumbed completely to German competition, the first Botany Bay wools arrived. Again to quote from Roberts: "Earliest consignments were a huge jest and men of the city gathered at Garraway's Coffee House in Cornhill to joke with each other about the new economic forces of the South and, not a little facetiously, to bid 10s. 4d. a pound for the first lone bale of convict wool that John Macarthur had shipped to the Dart, in 1807." Australian wool continued to come on the English market for six years without affecting its position relative to that from Germany or Spain. But about 1818 duties on colonial wools became nominal and freight rates dropped to 3d. per lb. remaining constant for years at that rate. By 1822 quality of Australian wool had so improved that Macarthur was awarded two gold medals for his exhibit by the Society of Arts. Roberts considers that to have been the turning point. Prohibitive duties, six times larger against German than against other wools, protected those from the colonies. Wool could be landed in England from Sydney or Hobart more cheaply than from Vienna or Leipzig.

Australian wool took the floor and held it, partly by its character, but largely by its price. Virgin land in Australia could be had for the occupation. English capital poured into the country. The whole of Australia was in a "ferment of sheep growing". Meanwhile in Germany "industrialization and the pressure of population made sheep growing a dwindling occupation.... No pasture land was left, the commons had been divided and the old right of depasturing stock on peasants' land existed no longer". No increase in sheep numbers was possible. Australia had the land, and it cost less for wool to go from Sydney to London than from the south of Germany. By 1840 Germany was beaten; in 1850 she was able to send England only a quarter of the amount that came from Australia. Eventually, the lead given by German firms when in 1845 they also imported Australian wools was followed widely. Our wools had completely ousted those of Europe from the English market.

But mark these dates. Sales were mounting from 1807 to 1840 and took a huge jump in the next decade. By 1850 Australia was supplying a little more than half the total wool received in Britain. During those years Australia's wool came from sheep that have practically disappeared. The two great strains of Australian Merinos, those of Murray and Peppin, were not established until 1843 and 1861 respectively. Early Australian wool was grown, on virgin country, relatively free from disease, and largely with any kind of sheep that squatters could secure. The greatest single factor in the growth of Australia's wool industry was price, and the outstanding ecological factor was the secondary environment established by economic conditions in Europe.

It is proper, therefore, to refer animal industries to requirements of local and oversea markets, the test being heritable adaptability of breeds and strains to those conditions with a satisfactory residual profit-margin for the producer. Costs of processes designed to modify primary environments with regard to disease, available nutriment, or other conditions, must be regarded as overhead in the industry and appraised accordingly. The first phase of animal breeding in Australia to be considered, therefore, is to what extent processes of animal husbandry are in accord with environmental conditions.

We have divided our continent broadly into areas for

crossbreds and British breeds of sheep, with fat-lamb raising as an important industry, and those in which Merino sheep predominate. We have no non-wool-producing sheep, nor have we a mohair industry based upon Angora goats. Our fine-wool sheep survive to the uttermost limits of our less arid areas and trespass upon our deserts. On the other hand, South Africa has a system of stratification which confines goats to the most arid areas, non-wool-producing sheep to those a little more favoured by rainfall and fine-wool sheep to the least arid of her grazing areas. Similarly America has a large Angora goat population in Texas and her other southern states west of the Gulf of Mexico.

In Australia, the same class of horse is found from the south to the north. The sugar fields of our tropical coastal belt on the east are cultivated by horses similar to those used within hearing of the Southern Ocean. We have no large number of asses and very few mules. In corresponding parts of Africa mules are common, and in the United States the proportion they bear to horses in each State is closely associated with latitude. For example, on a line from Pennsylvania on the east to San Francisco on the west coast, the proportion of all mules to all horses does not vary greatly at any point. The figures are: Pennsylvania 13·3, Ohio 6·6, Illinois 11·1, Iowa 6·4, Nebraska 2·0, Colorado 4·5, and California 15 per cent of mules. There has been no great preference for mules in the latitudes of these States. However, the position is different on a north-south line. The figures from the north are: Minnesota 1·5, Iowa 6·4, Missouri 26·1, Arkansas 60·5, and Louisiana 55 per cent of mules. In this direction choice has been definitely in favour of mules, particularly in the southern States. The possible inferences are either that we in Australia should be making more use of mules or that the American predilection for them is not founded on climatological data.

Our cattle industry has classified the continent into dairying and beef-producing areas. The former are confined to the coastal fringe and to irrigation areas. The remainder of the cattle country of the continent is stocked with British beef breeds and their crosses.

Beyond the dairying areas, particularly in Queensland, local conditions determine the use, in pastoral pursuits, of either fine-wool sheep or beef cattle. Thus on a line running west from Bowen on the east coast, there is first a belt of cattle country, then one of fine-wool sheep on the Downs, with cattle again towards the interior. Demarcations of this nature are controlled largely by the character of native grasses. When these are softer, with less penetrating seeds, sheep are depastured. In recently bygone years, sheep have been more profitable than cattle. This, combined with the political urge for closer settlement, has carried sheep into areas which previously were appropriated to beef cattle. Broadly, these are now depastured principally on areas considered to be unsuitable for sheep.

Regardless of their geographical situation, our cattle are of European origin. No incongruity is seen in stocking snow country of the Australian Alps with the same breeds that are used on Cape York Peninsula or in the neighbourhood of Darwin. Europe has no country similar to that north of the Tropic of Capricorn, and there are no grounds for the assumption that animals evolved in temperate climates are equally well adapted to tropical localities. However, this question has been fully discussed elsewhere (Kelley, 1943).

The foregoing broadly indicates the environmental associations of our animal industries. For the greater part they have been determined by expediency. Our European forefathers brought with them to this country the breeds and strains of animals with which they were familiar. With the exception of Merino and Polwarth sheep, of Corriedales imported from New Zealand, and of the New South Wales South Coast or Illawarra dairy cow, we have been content to populate the whole continent uniformly with the kinds of animals they bequeathed to us. There has been no careful planning; there have been few ecological studies, and only local or superficial vegetation or soil surveys.

All of these will be necessary to define fully the environmental limitations for our kinds of livestock. Until they are made, the existing associations between environment and animal populations in Australia must be regarded, for the greater part, as exploratory and largely fortuitous. However, there are areas on this continent where environmental conditions are completely favourable to their specific animal populations. The criterion for this is that the animals should continue, generation after generation, to reproduce themselves, without the introduction of new blood and without loss of vigour, vitality, or the characteristics of their ancestors.

Our Australian Merinos provide the outstanding instance of such a benign association between environment and animal requirements. Sheep such as the Murray Merinos, developed in 1843, and those developed in 1861 as the Peppin Merinos, are to this day satisfying these conditions on the properties upon which they were evolved. Using Lush's figure (1937) of about four and a half years for the average duration of a sheep generation, these periods represent 22 and 18 sheep generations respectively.

The production figures of such cows as the milking Shorthorn Melba 15th of Darbalara (32,522 lb. of milk, 1614 lb. butterfat in 365 days) in southern New South Wales, and of the Jersey Babs of Somerville (15,799 lb. milk, 843 lb. fat in 273 days) in Victoria, show that the environments of certain areas are suitable for dairying, and similar evidence could be presented for other industries and breeds of livestock. Nevertheless, problems still remain in our "marginal" areas and in others that do not have homoclimes of those in which the breeds we are using originated. However, to discuss these problems, except on broad general grounds, is beyond the range of this work, which, to maintain reasonable size, must assume that environmental conditions are sufficiently in accord with the needs of the animals to make progress possible by methods of animal breeding.

ANIMAL BREEDING

For this purpose, animal production in Australia can be regarded as controlled by extensive or intensive practices. Extensive methods are characteristic of large-scale pastoral activities where tens of thousands of cattle and even larger numbers of sheep are depastured in a semi-wild state with minimal control. Intensive methods are found on typical stud properties where matings are made deliberately, where pedigrees are kept and where animals are regarded as individuals with known and recorded life histories for each.

These are the two extremes, and there are many intermediate forms of procedure. Broadly, however, the two classes of animal management define the limits of practicability for improvement by animal breeding. Flocks in which it is impossible to identify the progeny of individual animals are necessarily restricted to mass selection on a phenotypic basis. When parents of both sexes and their progeny can be identified and when the performance of all can be measured, or appraised and recorded, such flocks or herds can benefit from genotypic selection. The difference between these two methods and the greater possibilities for improvement by the latter have been made abundantly clear. Animal breeding practices in Australia will, therefore, be discussed under the broad classification of extensive and intensive methods.

It should be realized, however, that size of the breeding project fixes no hard line between these two methods. Many small flocks and herds, in which intensive methods would be possible, are managed extensively. For example, it is reported that fewer than fourteen per cent of the dairy herds in any State of Australia are submitted for testing, that is, to measure the performance of individuals. Further, in many herds, the bull is considered to have fulfilled his function when the cows are in calf.

In many of our stud Merino flocks, intensive and extensive methods exist side by side. Nearly all have a super-stud group of highly selected sheep. Ewes of these are mated with special sires and, in some cases, the parents and progeny are identified and have their histories recorded. Very frequently the sire only is identified. In that case, pedigrees of the lambs are restricted to his side. Nevertheless, the method is sufficiently close to the required degree of intensity to enable genotypic selection to be practised. On the other hand, every large Merino stud has other component flocks. These are given different names such as select stud, double stud, or single stud, but all are extensively managed and based wholly on phenotypic selection.

The position is similar among beef cattle. There are herds which breed stud bulls from registered animals that can be identified and appraised. There are also other herds which breed herd bulls from cattle that lose their identity when the calves are weaned. These two groups may or may not be in the same ownership, but that does not affect the principle. It is that proposals to implement progeny-testing and genotypic selection may be sufficiently reasonable to receive serious consideration under one set of conditions, whereas they would be impracticable counsels of perfection under others. Nevertheless, as has been stated in earlier sections, the method of genotypic selection alone gives promise of any major uplift in production which properly could be attributed to animal breeding. If this is desired, then methods sufficiently intensive to permit genotypic selection must be used.

Extensive Methods

Reliance for improvement upon periodic culling and use of sires superior to the females are typical of well-conducted extensive production methods. The first has reference principally to the generation in which it is practised. Animals remaining after culling represent the general average, the good and the very good of a normal or quasi-normal distribution (p. 98). It has been shown that they can be expected to reproduce, in any generation, progeny not greatly unlike the populations from which they were selected. Any movement of the average for the new generation would depend wholly upon genes introduced by the sires. The methods are typical of those practised in our fine-wool flocks. In this connexion figures disclosing a steady increase in fleece-weights per head of sheep from about 5 lb. to an all-Australian average of about 8 lb. are familiar to all.

If an increase of 60 per cent in fleece-weight were a true picture of the effects of mass selection in a period of about 150 years, or approximately 60 sheep generations (Lush would state 33 generations), the rate of increase would be satisfactory. However, this is not the true picture. It does not give due weight to replacement of the 5 lb. type by another and much heavier producing kind of sheep. This took place either by rapidly grading up the 5 lb. sheep or more frequently by direct substitution. The idea of 60 per cent increase is based upon the yearly clip of early blade-shorn and "brook-washed" Merinos and a later day general average for the whole Australian wool clip. This now includes a relatively high proportion, over 20 per cent, of heavy crossbred and near crossbred fleeces. Further, it does not take into consideration the superior management, which, with other environmental factors, for example, clover burr, has played a large part in local increases that affect the general average.

The true position first has reference to the base line of 5 lb. *Dalgety's Wool Review 1926-7*, quotes Mr W. G. Brown, at one time sheep and wool expert of Queensland, as stating:

Nearly 80 years ago Frederick Bracken brought sheep to Queensland. The North British-Australasian Investment Coy stocked Rosenthal and Toolburra, in the Warwick district, with Saxon types. Comparatively small in the body, with superfine wool of short clothing type, their wool was heavy with yolk and had a heavy black shotty tip. The average fleece-weight was about 5 lb. greasy and $2\frac{1}{4}$ lb. scoured. These were typical of Early Darling

Downs sheep. Up till well on in the "70's" the fine, short wools prevailed. From that period they *have been replaced* in Queensland by the Rambouillet type with a good infusion of Tasmanian Merino. The Saxon type was utterly unsuited; the superfine wool developed into a cobweb and the animals were unable to withstand hardships of any kind. The introduction of Wanganella or sheep allied to the Peppin and Webber Rambouillet altered all that. The weight of the fleece was doubled and the sheep could withstand hardship.

Further evidence was put forward much more recently. Mr J. J. O'Carroll, who has been associated with the livestock industry for over half a century, wrote a reminiscent article in *Queensland Country Life* of 20 April 1945. He stated his first experience of classing and mating sheep was in 1893 on Jimbour when that flock was around 150,000 and there was also a large stud flock in which were bred all rams used on the owner-company's other properties. Up to a year or two before 1893 the flock had been Tasmanian. However, the sheep classer had been watching the Peppin evolution and induced the owners to turn to Wanganellas. Mr O'Carroll continues: "This man's opinion has been borne out by the success and profit that have come to graziers of the State, who, in the 90's and early years of this century, *discarded* the Tasmanian and Negretti strains and introduced Peppin, South Australian and Gamboola blood."

These statements of what happened in Queensland are applicable to other fine-wool-producing areas. Attention is drawn to the words "replaced" and "discarded" used in the quotations. It is evident the base line for comparison should not have reference to the sheep that were replaced or discarded but to other kinds. It is within them that breeding has taken place since "well on in the 70's".

It is generally acknowledged that the Peppin brothers developed their sheep from heterogeneous parents. On the sire's side principal importance is attached to a Rambouillet ram named Emperor, but there is a considerable amount of circumstantial evidence that Lincolns also played a part in the synthesis. Be that as it may, it was a very good piece of work, and there can be no doubt that fleece-weights of Peppin's sheep were considerably in excess of 5 lb. Further, their clean scoured yield was much above 50 per cent.

Importance attached to Rambouillet sheep in Australia about 1860 makes their parental flock of considerable interest. The original Rambouillet stud was established by the Government of France with 10 strains of Spanish sheep in 1786. In 1801 a further 40 ewes and 6 rams were brought from Spain. Since then none had been added up to the date of the report read by De Savignon who brought sheep from the stud to Australia about 1876. He stated that in 1813 the average fleece-weight of rams was 9 lb. 7¾ oz. and of ewes 7 lb. 13⅞ oz. In 1847 the figures were 12 lb. 2⅓ oz. for rams and 8 lb. 10½ oz. for ewes. In 1869 they were 14 lb. 10½ oz. for rams and 10 lb. 4$\frac{1}{12}$ oz. for ewes. These figures show that typical Rambouillet sheep at any time between 1813 and 1870 would be expected to clip well over 5 lb. of wool.

"Bruni" was the pen-name of a noted agricultural journalist in Australia in the latter half of the nineteenth century who was the author of two books, both now out of print, on sheep breeding in this country. He stated that the ram Lord Ripon, by Prince Imperial, left ram lambs which shore up to 21 lb. and ewes which cut up to 15 lb. 6 oz. In 1896 Lord Ripon's daughters are stated to have averaged 10 lb. 10 oz. Old Emperor, the Rambouillet ram of Peppin fame, is said to have cut a greasy fleece of 25 lb.; Premier 2nd, born 1881, a two-tooth fleece of 16 lb. 11 oz. and an adult fleece of over 21 lb.; and Invincible 2nd in 1885, at six years of age, to have cut 21 lb. 10 oz. The last three of these rams were of Peppin stock, two being used after Austin and Millear purchased the Peppin "double stud" flock in 1878. They should be compared with the eight-months' blade-shorn fleece-weight of three four-tooth rams selected from the reserve rams in 1942 to become sires in a very prominent stud where nothing is omitted from the scale of available nutriment. The weights were 20, 22 and 24 lb. with odd ounces in every case. The figures presented by Bruni do not suggest that 5 lb. is an appropriate base line for Australian Merinos, even after a reasonable discount is made for the fact that his figures were from stud sheep. He also gave 18 lb. 14 oz. as the average fleece-weight of stud-bred wethers about the same time.

Probably the true story of average fleece-weights among Australian sheep, as influenced by mass selection, is as follows. The early light-fleeced sorts, cutting in the order of 5 lb. fleeces, were eliminated. They were replaced principally by sheep drawn directly or indirectly from the Peppin or Murray flocks. These, at a very early date, were characterized by fleeces which averaged much closer to 7 or 8 lb. of greasy wool, and upon occasion were considerably heavier. A figure in the order of 7 or 8 lb. should be regarded as the true base line from the time that the early light cutters were replaced. The average fleece-weight per head of sheep in Australia is affected by the increasingly high proportion of crossbred and near crossbred fleeces and by seasonal variations. Nevertheless, the average for fine-wool station flocks has been maintained at or about the level of the original Peppin and Murray sheep. This is the result to be expected from mass selection as a method of genetical improvement.

This conception is supported by the following figures from a well managed and long established fine-wool station flock. They provide evidence as to the maximal effect possible from mass selection and its ultimate association with the curve of normal distribution (p. 111). The observations are stated descriptively, with approximate round figures, so that their source can remain anonymous. However, they are real observations. For many years the annual ewe-lamb drop of about 30,000 had been culled to 20,000 classified as first and second quality, 10,000 of each. The first quality were again classified into tops and average. There were thus three grades of ewes: about 1500 tops which would cut 12½ lb. of wool, about 8500 average which would cut about 9 lb., and 10,000 second quality ewes cutting about 8 lb. Over a number of years, ewes of these three kinds were maintained separately and mated with three grades of rams. The tops had the best available rams, represented by a 20-guinea grade; the average had the next best rams, represented by a 15-guinea grade; and the seconds had the next grade of ram represented by 13 guineas. These prices have reference to the general merit of the rams as judged by all aspects of their phenotypes. After a number of years under this system of breeding, progeny from each of the ewe groups gave the same weight of fleece. All had regressed to the mean of the population represented by that of the flocks in which the rams and the original 30,000 ewes were born.

Mass selection, as practised in all extensive activities, can be expected to maintain a level rather than to improve it greatly. In this respect it is of less service than an uplift in available nutriment.

Possibilities of utilizing hybrid vigour should not be overlooked in extensive practices. Different breeds of beef cattle are frequently crossed with good results. Thus Hereford, Devon, or Aberdeen-Angus bulls have been run with Shorthorn herds with considerable advantage. In one case, Aberdeen-Angus bulls were used until the herd was regarded as too close to Aberdeen-Angus type, when Shorthorn bulls were reintroduced. They in turn were replaced by Aberdeen-Angus when the herd became too concentrated on Shorthorns.

Experimental work in the United States by Phillips and others (1942) suggests the method to be well worthy of consideration by those producing commercial cattle. Its object was to investigate the possibilities of maintaining heterosis in beef cattle with a system of criss-cross matings. Hereford cows were the foundation and Shorthorn, Aberdeen-Angus, and Hereford bulls were the sires of successive generations. Data were presented for fifty-seven Hereford x Shorthorn steers compared with sixty-seven pure Herefords. Analysis showed the crossbred calves gained more rapidly in the feed lot than purebreds and were heavier at time of marketing. The percentage of carcass to live weight was greater in crossbreds which had fewer digestive disorders in the feed lot than the pure breds. In general the crossbred steers were less variable than the pure-bred steers.

These results are in keeping with those of other workers. Thus Shaw and MacEwan (1938) found that crossbred beef cattle had a definite advantage over pure-bred in rate of gain and quality of carcass. Of the various crosses tried by them, Shorthorn x Aberdeen-Angus carcasses gave the highest percentage of top grades.

This proposal is quite different from indiscriminate crossbreeding. It follows a preconceived plan designed to take full advantage of the hybrid vigour that results from heterozygosis within the herd. It will be recognized as one of the methods to be used, as two-way and three-way crosses, in the Bureau of Animal Industry proposals for cross-mating their inbred strains (p. 160).

A natural extension of the process is to mate more resistant with less resistant breeds and strains. In Australia the principal application of this method is for cattle in tropical environments. Crosses of British cattle with Zebus provide progeny repellent to ticks and with a hardihood that makes them better adapted to the low levels of nutriment associated with annual dry seasons.

Crossbreeding, and so hybrid vigour, is prominent among methods for commercial pig breeding. In fat-lamb raising also, two and three-way crosses, for example (long-wool x Merino) x Southdown, are used. Among Merinos, however, possible applications of the method are restricted to sheep of that breed, by the requirement of fine wool. There is, however, no apparent reason why, for extensive operations, a form of crossbreeding, really outcrossing, should not be practised with considerable benefit. Thus, when rams from a particular stud have been used until the breeding flock is strongly impressed by those "blood lines", a change is indicated to rams from another stud. The greatest probability of good results would occur when the two lots of rams were most different genetically. At least, for this purpose, rams should not be related to those used earlier. They would then also be unrelated to the earlier rams' progeny within the flock. The change would provide opportunities for compensatory matings by which the general character of the extensive flock could be modified. Unfortunately no figures are available to show results of this method.

Broadly, extensive operations will secure greatest improvement in production by heavier culling and by modifying the environment, particularly by increasing levels of available nutriment. Mass selection, which is, and probably must remain, the most widely used breeding method in this type of animal husbandry, tends rather to maintain polygenic hereditary characteristics at or about their original levels. Within the method the only possibility of raising these resides in the superior nature and degree of character fixation, in the genetical sense, of the introduced sires. When the sires are homozygous, that is, prepotent for specific characters, as the result of the breeding method by which they were evolved, movement of the mean will be maximal in those directions. Maintenance of high degrees of vigour will be possible by well-planned outcrosses. These may be between widely different kinds, such as Zebu and British cattle, between different breeds, such as Aberdeen-Angus and Shorthorn, or, as in fine-wool flocks, between different strains, for example, mating progeny by rams from one long-established Merino stud with rams from another strain. Because of their limitations on improvement by animal breeding processes, extensive methods should not be practised when those more intensive are economically possible. Intensive methods should be applied, wholly or in part, wherever it is possible to do so under practical conditions or some reasonable modification of them.

Intensive Methods

Any of the methods of animal breeding discussed in the preceding chapter are applicable in Australia when, as in stud breeding, matings are controlled and progeny are identifiable. It should be realized, however, that in the

United States it is recognized some of the methods are too heroic for private enterprise. Professor Wilson, when discussing the cattle-breeding project at Davis, California, was quite definite work of that nature should properly be undertaken by institutions. His reasons were that the processes were too expensive and that good results were too problematical for such projects to be undertaken by private owners. This is regarded as a very sound opinion.

In some States of Australia ordinary stud herds and flocks in Government ownership conform to and compete with normal stud practices of private ownership. This would have most merit if the quality of the stock were so high that they could properly be regarded as nuclear groups evolved according to Hagedoorn's conception (p. 169). It would have virtue also if relatively large sums of public money were expended to purchase overseas outstanding proven sires at prices beyond the range of private persons. These animals, predominantly males, could then be made available to accredited herds by artificial insemination. It is thought that the resources available for all aspects of Government enterprises and the co-operation that could be extended by private breeders who would benefit from such a plan, together could make an outstanding contribution to animal production in Australia.

Such flocks and herds could be used to practise oversea methods of analysing and classifying animals according to their genotypic value. There can be no reasonable doubt that, within the breeds and strains now in Australia, some animals are the equal of any in the world. However, those with superior germ-plasm require discovery, fixation, and propagation. The proper function of State flocks and herds would be to provide nuclear groups unprocurable by methods in accordance with private ownership. If such outstandingly superior strains became available to private stud-masters, they would undoubtedly make full use of them by normal practices.

(a) Cattle

Progeny-testing and evaluation of dairy bulls by the production of their daughters involve herd-testing. These two methods provide all that is necessary in the way of animal breeding to uplift the dairy industry considerably. With appropriate feeding, they will do in Australia what they have done elsewhere. For example, one Victorian herd of 23 cows, in 1942, averaged 530 lb. of butterfat without allowances for the age of individuals. The high production of this herd resulted from using three proven bulls in sequence. They were father, son, and grandson, respectively.

There is a trend in dairying taking place in England and America that is worthy of note. In these countries human populations are relatively dense and beef prices are high. Cows are regarded as potential carcasses. The milk and butterfat they produce are by-products. This emphasizes the dual-purpose cow and depreciates true dairy cattle, particularly Jerseys. However, in Australia, dairying is a primary industry and, within it, the methods indicated are sufficient to improve production.

The problem of breeding beef cattle is to find a measure of excellence by which animals can be progeny-tested. There is, of course, the normal procedure of show-yard evaluation. This has some merit, but there are many instances where judgments of cattle on the hoof are contradicted by appraisals of their carcasses.

Absolute merit in a beef beast is associated with rate of gain per unit of food intake. Animals that reach market weights at early ages are better economically than those which produce similar weights of carcass at greater ages. For example, a trial was made in Australia by sending for slaughter lines of two-, three-, and four-year-old bullocks and steers. There were three- and four-year-olds of one kind, and two- and three-year-olds of another. Gross returns from the lot of three- and four-year-olds were £7 3s. 8d.; from the three-year-olds of the other kind they were £9 8s. 3d. and from the two-year-olds of the second kind they were £7 7s. 11d. However, the £7 7s. 11d. was better than the £7 3s. 8d. by much more than 4s. 3d. The latter was made at the rate of £2 6s. 3d. per bullock year. Three-year-olds of the second kind earned £3 2s. 9d. per bullock year, but the two-year-olds earned at the rate of £3 13s. 11d. per bullock year. The two-year-olds produced at the annual rate of 301 lb. of carcass beef in their two years, but, as judged by the weights of their three-year-old mates, they would have produced only 151 lb. during a third year.

Criteria of this nature can be secured only after the animal is slaughtered, and to suggest that representative progeny of pedigree beef beasts should be appraised in this manner appears to contradict the aim of pedigree breeding. Nevertheless, these are the true criteria of excellence. It is stated by Youatt that Bakewell kept the hall of his home as a museum.

> The separate joints and points of each of the more celebrated of his cattle were preserved in pickle or hung up side by side; showing the thickness of the flesh and external fat on each and the smallness of the offal. There were also skeletons of the different breeds, that they might be compared with each other, and the comparative difference marked. Some joints of beef, the relics of old Comely, the mother of the stock who was slaughtered when her existence had become burdensome to her, were particularly remarked. The fat of the sirloin on the outside was four inches in thickness.

Slaughter of some of the progeny is necessary and unavoidable. Further, those slaughtered should be not culls but representative samples of the progeny. However, the cows used to test bulls need not be valuable registered animals. It has been stated by the U.S.A. Division of Animal Husbandry that rate of gain in live weight per unit of food ingested provides guidance for progeny testing. This at least should be determined in any b[illegible] stud. Criteria of this nature provide indications of excellence far and away more realistic than, for example, the nature of the coat. With regard to this, the following is an extract from notes on preparing beef cattle for exhibition that appeared recently in the country press of an Australian city.

> The calf should be kept in a cool barn during the day and allowed to graze at nights for a few months prior to the show. The hot sun . . . will sunburn the hair and cause it to appear somewhat dead, rough, and coarse, and to lose its glossy appearance. . . . The stable also may be darkened and many flies kept out. . . . A few weeks before the show the calf may be blanketed to advantage. The blanket . . . is used to keep flies from worrying the calf, to give the hair a more glossy appearance and to keep mellow or soften the hide. The blanket for this purpose should extend from the neck to the tail and come down over the sides.

The calf, judged under these conditions, is supposed, by these criteria, to be more or less suitable to beget progeny which, together with a very high proportion of the beef cattle of Australia, will be depastured under natural conditions within the tropics. There they will be assailed periodically by mosquitoes, sandflies, and buffalo flies. In comparison with these, a few stable flies will only be remembered as a source of entertainment.

Criteria for selection must be realistic and used in progeny-testing. Superior animals so isolated can be used as is indicated in the previous chapter. The criteria for beef cattle are not principally those of milk-yield and butterfat production. However, early maturity, body form, and all other worthwhile attributes of beef cattle are similarly polygenic and are amenable to the processes discussed for controlling such characters. They are equally applicable to mutton breeds of sheep.

(*b*) *Fine-wool Sheep*

Evolution and maintenance of fine-wool sheep among which rams yield 20-30 lb. of wool and large numbers of ewes 16 lb. is an outstanding achievement by any criterion of judgment. It has required exploitation of the crafts of breeding and feeding, and probably no process in Australia is more worthy of an analysis of methods.

The general arrangement of stud Merino breeding in Australia is very characteristic. There are the "foundation" groups. Stud sheep descended from these are divided into sub-groups that have been isolated from outside blood for a number of years. In this respect they are similar to those of Hereford cattle isolated by colour faddists in the nineteenth century (p. 50). Frequently there are large "parental" groups, and those with greatest repute have given rise to many "daughter" studs. It is common, therefore, to see more recent flocks advertising

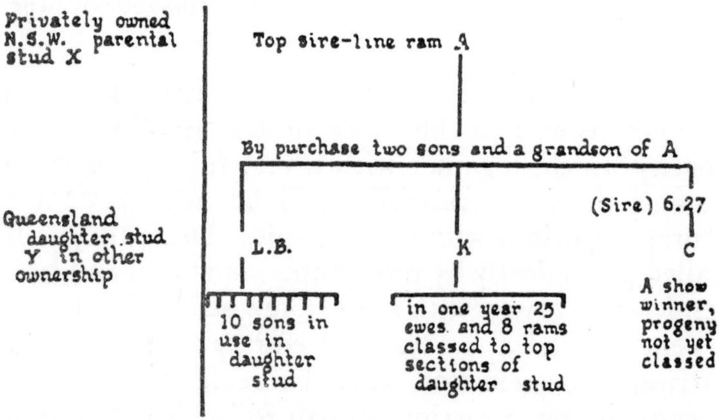

FIG. 45.

Diagram showing transposition of principal sire-line of parental stud to daughter stud in sheep.

that they are founded upon the "blood" of particular studs which, in turn, claim their own descent from such foundation flocks as those of Peppin or Murray.

Sometimes the association of the daughter studs is not very close, but at other times it is designed to transpose the principal sire-line of the parental stud. For example, a recent announcement gave sufficient information to enable the accompanying diagram (Fig. 45) to be prepared.

The diagram shows that the owner of the daughter stud, Y, has repeatedly sampled, apparently with satisfactory results, the genotype of the ram A. He has therefore elected to build the gene aggregate of his own stud flock upon that basis.

This could have been done by artificial insemination, ram A being the donor, had suitable arrangements been possible. His genotype would then have been sampled "undiluted". If the proposal had been made it would probably have been rejected. Herein is the biggest limiting factor for artificial insemination in this country as distinct from Russia. There, such an animal as A would be owned by the community. Here, he is owned by a private individual who secures greater revenue by selling descendants at up to, say, 670 guineas, the price paid for the grandson C, than he would derive from semen samples.

The method used in the daughter stud incorporates with each sample of the genotype of A another, derived from the dams of the purchased sires. The "dilution" is increased in the ten sons of L.B., the eight sons of K, and in C.

Advisability of concentrating upon the genotype of the ram A can only be determined by the uniform quality of his progeny. However, it is established in the mind of the owner of Y. This case, as an alternative to artificial insemination, suggests the use of judicious inbreeding. This could be achieved by mating the better or possibly both of the purchased sons with their own daughters, if their suitability for the purpose were established by the merit of their earlier progeny. The method would bring the concentration closer to the genotype of A, and restrict its dilution, which is inevitable as A becomes progressively a more remote ancestor.

The third possibility is the method, used by the owner of Y, of making repeated samples or near-samples of the genotype of A; in this case two samples and at least nineteen near-samples. This, as has been shown, introduces diluents; however, it reduces the risks of inbreeding. The position is interesting because it discloses three possible methods to achieve a similar end, that is, concentration upon ram A in the daughter stud.

Excellence as displayed by selected sheep in shows and sales undoubtedly depends largely upon optimal environmental conditions. It cannot be wholly accidental that most of the Merino studs are in country where heavily nitrogenous annual-clover growths, with residual seeds, occur periodically. In addition to these, abundant food is provided for young rams and particularly for those reserved as two-tooths for subsequent reappraisal.

There are two schools of thought with regard to procedure of this kind. The first states that unless the most favourable conditions are provided, animals are unable to demonstrate their full hereditary potential. Those that respond are most suitable for reproduction (p. 178). The opposing school does not contest this point, which is well established as an indication of probable excellence. It suggests, however, that the environmental conditions provided should be realistically controlled; that they should at least approximate those in which the progeny of the animal will develop. The conception depreciates unusual performances, in this case wool growth, and classifies all as equivalent that are obviously outstanding.

When the second view is referred to the curve of normal distribution, all such performances lie in its highest segment. They are within the area (Fig. 33) enclosed by the end of the curve, the base between $+2\sigma$ and $+3\sigma$, and a perpendicular erected from the base at the point marked $+2\sigma$. This conception is valid unless it can be shown that similar high orders of merit characterize *all sibs* of a family. But it is inconceivable that even a very small number of animals could be homozygous for so many characters and thus uniformly beget such progeny. There are too many genes involved. The position is similar for any extreme production records because the characters upon which these depend are all polygenic. Nevertheless, the keen competition of vendors at shows and sales and the uninformed condition of the buying public justify the practice of high feeding, at least as a good business method.

It has been observed, however, that stud-masters usually retain for their own use the best looking of the reserve rams and that these also have been highly fed. It is evident, therefore, they subscribe to the thesis put forward by the first school of thought. Selection of this kind is wholly phenotypic. The nature and possibilities of this have been discussed and it has been emphasized repeatedly that genotypic selection is an incomparably surer method. The question arises, how could the method be practised? It would possibly be unwise and impracticable to suggest that many of the reserve rams, at about fifteen months of age, should be adequately progeny-tested by normal methods of mating. However, it is suggested that this could be done with the aid of artificial insemination. Further, in the sheep breeding experiments described in the U.S.A. (p. 164) ram lambs are progeny-tested when 7 or 8 months of age. Eight of these were observed by the author to beget up to 100 per cent of lambs and never less than 60 per cent from lines of 20 test ewes under normal conditions of mating.

Pedigrees, at least on the male side, are available in the majority of studs. When reserve rams are isolated by all ordinary methods of selection, it should be possible, by an examination of their pedigrees, and with the information these reveal to the stud-master, to narrow the range of possibilities for future sires to a relatively small number of sheep. It would not hurt these to become sires, by artificial insemination, of progeny numerous enough to determine which were valuable sires. The proposal to use them as two-tooths would probably involve withholding the small number of rams from sale until they were approximately three years of age. Against this can be offset a reasonable degree of certainty that the best sires in any generation would be retained for use within the stud. Further, because the test could be made with any reasonably good uniform line of ewes, the breeding life of the highly selected ewes in the top stud would be conserved for use with sires already proven. There is also the possibility that, by use for test at still lower ages as is done in the U.S.A., the rams could be progeny-tested before sale as four-tooths.

Most large stud flocks are divided into sections. The most highly valued of these will be called the top stud. It contains a number of ewes and a small number of sires. Ewes in this flock are usually discovered to be on heat by vasectomized or "teaser" rams, and are mated once only under controlled conditions with one or other of the sires.

A number of interesting points arise from this procedure. The first is the degree of consanguinity that characterizes the matings. Advantages and disadvantages of close breeding have been discussed (p. 141), and will not be repeated. The question arises, however, as to the data that are available to determine the degree of inbreeding practised. There can be no doubt that a judicious degree of inbreeding will always be an important method in stud breeding. It is highly desirable, however, that it should be deliberate with regard to both its direction and its intensity. These can only be determined when "two-sided" pedigrees are available for both parents.

It is a common practice for matings to be made between descendants of this and of that ram. The descendants are spoken of as being of this or of that "family", having reference to rams that founded the strains. Subdivision of the top stud into such groups is very sound, particularly when, as frequently happens, some lines are stronger than others in specific points. Indeed this aspect of sire-line selection should be emphasized, because the fewer the number of characters for which selection is being practised in a line at one time, the greater is the chance of success. However, the point being made is, when there are two, three, or four sire-lines and it is the policy of the stud to cross these, a much higher degree of control can be exercised when pedigrees are also kept on the ewe side.

Fig. 46 shows the pedigree of a stud ram No. 27, abstracted from the records of a flock in which he was valued highly. Numbers have been used to replace names and secure anonymity. The pedigree shows two families being used. These, for convenience, have been called the No. 2 and the No. 3 lines. Concentration of No. 27 on No. 3, by the data shown, is only 0·39 per cent, and none is shown to No. 2. It is known, however, that, in the stud, when the two rams No. 2 and No. 3 were alive and equally outstanding, it was the policy to mate sons of one with daughters of the other. Probably, therefore, if it were completely known, that generation of the pedigree would appear as is shown in Fig. 47. It now conforms to

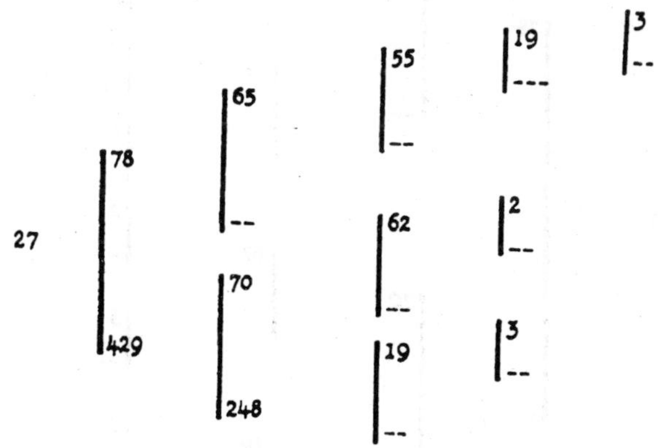

FIG. 46.
Pedigree of ram No. 27.

what was the stud procedure as disclosed by the records. No numbers other than 2 or 3 have been added to reconstruct the pedigree, but it is highly probable that the blank for the maternal grandsire of 78 would have been No. 19. However, by taking into consideration only the repetition of No. 2 and No. 3, it is now learnt that No. 27 was concentrated 2·34 per cent on No. 3 and 3·13 on No. 2. The coefficient of inbreeding of No. 27, from these data, is 6·64 per cent. Increase over the sum of 2·34 per cent twice is due to the closer reappearance of No. 19. To establish the point being made with regard to No. 2 and No. 3, this was not considered previously.

The difference between 0·39 per cent and 6·64 per cent is appreciable. No significant increase in uniformity could be expected from the first, but the second is approximately equivalent to a mating of first cousins, which gives 6·25 per cent as a coefficient of inbreeding. The first could be increased without much thought for the consequences, but any increase over the second may or

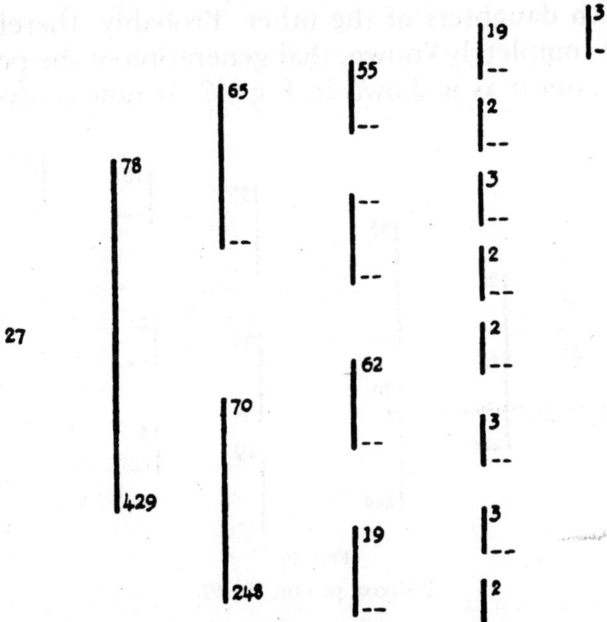

Fig. 47.
Reconstructed pedigree of ram No. 27.

may not require consideration. This was referred to when discussing test matings in controlled heterosis (p. 170). It is interesting also to compare the reconstructed pedigree (Fig. 47) with the use Booth made of the Shorthorn bulls Leonard and Buckingham as shown in Fig. 8.

Another advantage to be derived from ewe pedigrees is that the records will then disclose degrees of fertility that characterize the ewes. Fertility is the ability of a ewe to bear progeny, and a highly fertile ewe is one which bears more progeny than the average. The character is not the same as twin-bearing. Twinning is largely controlled by age and by environmental conditions prior to mating, but there may be an inherited tendency to bear twins. When highly fertile ewes also produce twins they are prolific. However, fertility is the character under consideration. When no ewe records are kept it is difficult to isolate ewes that bear lambs but do not rear them, or ewes that are occasionally fertile. Completely sterile ewes may be discovered from their general appearance, but those that are occasionally fertile although "shy breeders" are not equally obvious. Further, when breeding ewes are being classed upon their appearance these inferior mothers always "take the classer's eye" for retention, and frequently the culls include many of the regular breeders.

Observation has also shown some Merino strains reach sexual maturity slowly. Such ewes may bear 23 per cent of lambs as two-tooths, 53 per cent one year later and 78 to 82 per cent when three and four years of age and older. Compared with cross-bred ewes under the same conditions the earliest percentage is in the order of 50 per cent lower.

Insufficient data have been accumulated to prove the hypothesis, but the evidence strongly suggests levels of fertility to be hereditary. Probably degrees of fertility depend ultimately upon an inherited sufficiency or deficiency of secretions from the anterior pituitary gland. There can be no doubt, however, that infertility has on many occasions been associated with inbreeding. Because every stud-master rightly desires some degree of uniformity, he must practise inbreeding. It follows, therefore, that he must also be on watch for all its known sequelae. Adequate care in this respect can be exercised only when sufficient data are available for consideration.

Merino ewes are generally indifferent milkers. Experimental observations disclose their daily yield is seldom greater than 1400 grammes and that it may be as low as 680 grammes. This should be compared with a yield of 2000 grammes by British-bred ewes in the same period of lactation. There is no reason to suspect that milk production in ewes is inherited in a manner different from that in cows. The character has been discussed for those animals and the discussion is transferable to sheep. Having in mind the importance of adequate nutriment in the early stages of post-natal life, selection should be made among Merino ewes for those that are "good mothers". This is another reason why records of performances of ewes should be kept.

The only justification for discarding ewe records is that the progeny of rams greatly outnumber those of individual ewes. Although it is relatively easy to apply a common ear-mark or otherwise permanently to identify the progeny of particular rams that are individually mated, those of each ewe must be ear-tagged or tattooed separately for a similar purpose. However, when either of these is impracticable, there are systems of punch-holes and ear-notches by which numbers can be given to individuals. If one of these methods were used in the top stud at lambing time and a note made of the corresponding number of the ewe, the additional clerical work would be slight. It should be recognized that the more extensive the recorded data become, the greater is the volume of material available to improve selection.

Top studs bear a fairly general relation to the remaining portions of the stud flock. The general position is that each year the young sheep are classified phenotypically. Some of the ewes are sent to the single stud, there to replace older ewes. A better grade of ewe goes to the double stud, and the most excellent to the top stud. Sometimes there are outstanding young ewes among progeny from the double and single studs. When these occur and are sufficiently excellent, they may go into the top stud.

Although progeny of the stud as a whole, these ewes do not belong to any of the recognized families in the top stud, nor have they pedigrees. The older stud overseers resent these interlopers and frequently maintain that they should not be moved into the top stud. However, the procedure has advantages. The ewes may be closely related to animals already in the top stud. When this is so they cannot reduce the general level of inbreeding even if they produce a lamb that eventually becomes a top stud sire. However, the chances of their being of this nature become progressively more remote as the size of the two remaining sections increases. Generally they can be regarded as something of an outcross. Their use is analogous to the method previously discussed as characteristic of the Booth family of Shorthorn breeders (p. 21). They correspond also to the animals introduced periodically under the scheme of controlled heterosis. Their principal disadvantage is that their selection is phenotypic. Were they to be introduced later, when known to have been dams of excellent lambs, the move would be very sound. The question as to whether sires of similarly unknown origin should be used in the top stud also arises frequently, and the same comments are applicable.

The general character of breeding within top studs varies, but frequently it is similar to that described as controlled heterosis. The possibility of isolating and fixing desirable genotypes by recognized methods has been discussed as more suitable for the flocks of institutions than for those of private owners. This, of course, has reference to heroic degrees of inbreeding and to the risk of loss associated with them. It does not refer to judicious inbreeding. Generally, top studs closely approximate the ideals of intensive methods of breeding. On the other hand, mass selection controls operations in the other stud sections. If these two processes were equally advantageous methods of breeding, the proportions of reserve rams classed from the various sections should be in the order of their numerical strength. The association could be broken only by the higher degree of selection and by the more specific criteria for matings in the top stud; parentage is at random in "paddock matings". That the numerical association does break down is shown by recent figures. These disclosed that sixty per cent of the reserve stud rams in a particular year came from top sections in which controlled matings were practised. The remaining forty per cent came from the other sections. This was so despite the fact that the top section contained only approximately one-seventh as many ewes as the remainder.

Broadly, the less highly selected and more extensively operated sections of studs provide flock rams for pastoralists. The methods of random mating and of mass selection by which they are bred are unlikely to result in any marked degree of homozygosis. The usefulness of rams bred by these methods, therefore, depends upon their position in the curve of normal distribution relative to that of the sheep with which they will be mated. This is critical with regard to either maintenance or improvement of the commercial flocks. The difference between the average merit, determined before classing, of the flock in which the rams are bred and the average, determined similarly, of that in which they are used, is the only possible measure of their influence.

The problem for commercial wool-growers, therefore, is how to justify economically the use of more intensively bred top stud rams. On the one hand there is the probability of considerable improvement in their flocks. On the other there is the certainty of greater outlay for the more soundly bred rams. Further, the outlay would increase if purchasers were to insist that stud-masters present to them evidence of genotypic selection to establish values.

Mass artificial insemination may provide the answer. Although it would require outstanding organization and considerable expense to practise the method on a large scale, it might be possible and advantageous for commercial wool-growers to isolate a section of their large flocks for the purpose. Whether this can be done or whether there is any economically sound possibility of commercial flocks being mated extensively with highly bred progeny-tested sires is problematical. On the other hand it should be possible within studs so to tighten up procedure that progeny-tested sires are used, possibly by artificial insemination, at least in isolated portions of the lower-graded stud sections. If this were done for several consecutive generations the encircling chains of mass selection would be broken. Real progress, by breeding as distinct from feeding, is possible only by some such method. Without it characterization must continue to swing between the means of the two populations involved. With it the movement would be controlled by the genotypes of particular individuals.

CHAPTER XIV

ARTIFICIAL INSEMINATION

PLANNED use of artificial insemination on a grand scale in Russia has drawn considerable attention to the process in recent years. It is not new, but it would be difficult to determine how old it is. It is said to have been used by desert Arabs about 1300 A.D. (700 of the hegira). It is stated also that the method was crudely employed by Arab tribesmen who, after a mare was served, transferred a sponge from her to another. However, the first authentic account of its use in mammals is stated by Lambert and McKenzie (1940) to have been in 1780 when an Italian named Spallanzani was successful in artificially inseminating a bitch. Between 1884 and 1896 similar experiments were repeated in England. Soon after the beginning of the present century potentialities of the method began to be well recognized. In 1909, a Russian named Ivanoff, working under the old regime, reported successful experiments, and in that year a laboratory was established for the purpose of training veterinarians in the technique. These studies and experiments were interrupted by the first World War and the Russian revolution. However, they were again undertaken to such good effect that in 1936 Kersin reported upwards of 6,000,000 sheep and cattle had been artificially inseminated in Soviet Russia in that year.

Investigations have also been undertaken in England, the United States, and elsewhere until a very large volume of information has been collected. The general trend of experimental procedure has been to improve collection of semen, to evolve methods for handling it, and to develop optimal procedure for the actual insemination. All of these processes have required an extensive knowledge of the physiology of reproduction in females, particularly with reference to the time and cause of oestrus and ovulation, and of the biology of sperm-cells. These two are critical because, whether the insemination be natural or artificial, reproduction will take place only if a live spermatozoon meets and fuses with a viable ovum.

Under normal conditions, the time of coitus, in which insemination is the essential feature, is fixed by the female of most species of domestic animals. Exceptions are camels, deer, and certain others, in which the male periodically enters a season of sexual activity. Duration of oestrus or heat varies. Probably, among ordinary domestic animals, the bitch has the longest period. In any case, it is so obvious that it lends itself to description as a typical instance of the three phases normal in all females: coming on, on, and going off. Phenomena of heat are caused by absorption of fluids which eventually escape from a ruptured vesicle in the ovary. That vesicle contains the ovum or egg which is extruded into tubes leading to the uterus (breeding bag). Ordinarily every offspring represents one ovum, but there are relatively rare instances when one ovum gives rise to two individuals, referred to as identical twins.

After leaving the ovary the ovum travels slowly down the tubes and, if it is not fertilized in them, passes to the exterior. After extruding the ovum the ovary becomes quiescent for a period referred to as one of anoestrus. This is of varying duration. In the bitch it is usually about six months, but it may last twelve. In the breeding season of sheep the time between heats is called a period of dioestrus to distinguish it from the longer non-breeding or anoestrous season. However, the cause of both periods is the same, that is, quiescence of the ovary. Average duration of dioestrus in the sheep is seventeen days; in cattle it is twenty-one days; and in mares, after the foal heat, approximately the same as in cattle.

Extrusion of the egg or eggs is called ovulation. It usually takes place late in oestrus or immediately after the female ceases to accept the male. These facts are critical alike for artificial insemination and for controlled or hand service as practised in studs. Inseminations made at a time when ovulation cannot be expected, as could happen in artificial insemination, or at times too early in oestrus, as does happen in normal mating, would be too remote from ovulation. They would therefore be infertile. The reason in the latter instance is that sperms do not live long in the female genitalia. If service is too early in a prolonged oestrus, the sperms will either be dead or devitalized and so unable to fuse with the ovum when it is extruded. For this reason, services under hand-mating conditions are best deferred until late in heat. This is equally true for artificial insemination.

There is no special period for production of sperms by most male domestic animals. They are continuously produced in the testes and held stored ready for emission when coitus takes place. They have a much longer life in the male genitalia than in that of the female; nevertheless, they do die and are absorbed, their place being taken by others. Fluids in which sperms are suspended are secreted by what are known as accessory sex glands; the combination of fluids and sperms constitutes semen. In addition to providing a vehicle for sperms, the fluids activate them and cause them to exhaust their store of energy. They then cease to be motile and thus have their fertilizing power reduced or neutralized. For this reason a very essential part of work that has been done in elaborating techniques for artificial insemination has been directed toward conserving life and energy of sperms.

BIOLOGY OF SPERMS

Mature male egg-cells or sperms are minute. So minute that two drops (2 minims) of a normal ram's semen contain sufficient individual sperms to produce thirty million more than all the sheep in Australia if each sperm could be isolated and caused to fertilize one ovum.

Sperms have a minute body or head and a thin tail approximately four times as long as the head. When a fresh specimen of semen is examined under the microscope, the sperms look like an immense number of very small tadpoles swimming through the field of vision. Movement of their long tails drives them forward with a rotating corkscrew motion, due to body-shape.

Their motility is critical for fertility because, when introduced to the female genitalia, either normally or artificially, sperms must ascend the whole of the genital tract in order to reach the ovum. Whether they could or do travel the whole distance under their own steam, as it were, is still a moot point. It is reasonably certain, however, that the upward passage is greatly assisted by contractions of the female genitalia and by the ebb and flow of their mucous secretions which carry sperms hither and thither and sometimes in an upward direction.

This indiscriminate movement gives rise to enormous wastage and the number of sperms is proportionally large to offset it. The point is of great importance because it restricts the degree of dilution that can be practised and so limits the number of artificial inseminations that can be made from an individual ejaculum (emission).

Intense activity consumes the stored energy available to sperms and shortens their lives. Their movement is associated with the breakdown of sugars (glycolysis) and can occur in the absence of oxygen, provided glucose is present. In this respect the metabolic activity of the spermatozoon is similar to that of ordinary muscle in which the up-take of oxygen is associated, not with the contraction, but with oxidation of secondary products. Oxygen consumption is therefore not a direct measure of sperm activity, but it runs parallel with it. Consumption of oxygen by sperms is measurable and the respiration rate of sperms is determined from it. Fertility, assessed by the average number of normal services by bulls for each conception, has been shown to be related in particular semen samples to the respiration rate of the sperms. Walton and Edwards, of Cambridge, give the following figures which show the relationship.

TABLE XIV

FERTILITY AND THE RESPIRATORY RATE OF SPERMS

Number of bulls	*Services per conception*	*Average initial respiratory rate*	*Average respiratory rate at end of two hours*
3	1 – 1·9	·286	·058
6	2 – 2·9	·187	·035
4	3 – 6·9	·127	·012

Respiration of sperms is only one of their characteristics. Degrees of fertility have been associated also with abnormal shapes, with insufficient concentrations and with a number of other attributes. All of these are broadly referred to as pertaining to or comprising the biology of sperms. Respiration rate was mentioned specifically because it is among the newest of the criteria of excellence and offers considerable promise as a method of valuing semen. Such tests are most necessary, because the fertilizing power of semen resides only in the spermatozoa it contains and they are very delicate cells. Temperatures of 114·8°F. will kill them although 101·5°F., only about 13°F. lower, is the normal temperature of many domestic animals. Curiously, sperms appear tolerant of low temperatures. These, in effect, put them to sleep—like apples in cold storage. More will be said later of this attribute. In it lies the hope of transporting viable sperms over long distances. By it, progeny have already been secured in Poland from semen collected in England, and in Brazil from semen collected and dispatched from the neighbourhood of Washington, U.S.A.

It is customary and proper to examine semen after collection. The points considered are (i) volume, (ii) colour and cloudiness, (iii) consistency, (iv) motility of sperms,

(v) the number of normal sperms per cubic millimetre of semen, (vi) freedom from bacteria or cells that indicate disease conditions of the male genitalia, (vii) the degree of acidity or alkalinity (hydrogen-ion concentration), and (viii) the respiratory rate. In well-ordered associations for the extensive use of males by artificial insemination, semen examinations are complete in all of these respects because they are related to fertility.

MANIPULATION OF SEMEN

Handling semen after collection is a critical phase of artificial insemination.

The method to be employed depends on the time that will elapse between collection and insemination and on the species of the donor. If the semen is to be used within two hours of collection, the technique is relatively simple. However, such speedy use is incompatible with inseminations to be made at long distances from the point of collection, and use of the process for such work is one of its outstanding possibilities. Accordingly, a large volume of research is taking place to maintain the life of sperms outside the male organs.

Sperms in the genital organs of rams, for instance, have about sixty days of life. As judged by their motility, sperms have been kept alive after emission for the same period. This was an extreme case, however, and there is no evidence to show that the semen, though retaining motility, was also fertile. The two characteristics are not necessarily equivalent. A more common period of life under laboratory conditions is from fifteen to twenty days. Two successful impregnations have been recorded by Winters, an American worker, with ram semen kept in the laboratory for six days, three with semen kept five days, several with semen three days old and many with one- and two-day-old samples of semen. Dr Gunn of Sydney University has reported as high a percentage of impregnation in ewes with ejacula stored twenty-eight hours at 6°C. as with comparatively fresh ejacula used four to six hours after collection. Broadly, however, although a few successful impregnations have been reported with transported semen used up to ninety-eight hours after collection, three days should be regarded as the extreme upper limit for probable success in our present state of knowledge.

Methods to prolong longevity of sperms are aimed principally at the removal of accessory fluids from the semen, particularly when these predominate, as in the ejaculum of the horse and the boar, and at the exclusion of light and air. These are in addition to lowering the temperature. In Russia one of the newer methods is to store spermatozoa in gelatine. It is the work of Milovanov. While still warm, semen is mixed with gelatine and then run into small tubes where it sets when cool. The small suppositories so formed are transported to the female to be inseminated. When inserted in the opening to the neck of the uterus (cervix) the gelatine melts and sets the sperms free.

Accessory fluids are removed either by centrifugalizing the semen, or, as is readily possible when semen is obtained by electrical stimulation or by masturbation, by not collecting the fraction of the ejaculum that is composed largely of such fluids. Special diluents have been found advantageous to replace accessory fluids and prolong life in some forms. The diluents have to be made according to very accurate prescriptions, which quite frequently call for the use of sodium and potassium hypophosphate in addition to calcium sulphate. Sometimes the yolk of fresh eggs is mixed with the first two of these and sterile distilled water. Diluents containing egg-yolk have given particularly good results with semen of bulls. However, no good purpose would be served by elaborating further on such techniques. The important points are that research, designed to prolong the fertilizing life of spermatozoa after ejaculation, is taking place along such lines and that the hope of long distance artificial insemination depends upon its success. At present the process can be regarded as restricted to points probably not farther apart than a distance equivalent to about thirty hours of travelling time.

For times of that order, it is necessary for the semen of the bull, ram, or goat, to be collected with great care and kept at a suitably low temperature. A United States Department of Agriculture publication states that, for such a journey, the semen is collected in a vial and covered with a layer of high-grade neutral paraffin oil up to the cork. This eliminates any air space in the vial. It is then wrapped in cotton wool and placed in a screw-topped watertight glass vial which is well protected by further layers of cotton wool held in place by rubber bands. The package is then placed in a quart vacuum bottle and tightly packed with chipped ice but no salt. The whole may now be transported, but the risk of breakage is reduced if the vacuum bottle has a top and bottom cotton wool pad.

When the semen is removed from the vacuum bottle, its temperature must be raised gradually. This is done by allowing the containing vial to stand at room temperature for about thirty minutes after which it is stood in tepid water (90-95°F.) for a few minutes. Care should be taken to avoid temperatures above 95°F.

When inseminations are to be made almost immediately after collection, as happens when the process is being used in flocks or herds adjacent to the donor, elaborate procedure is not necessary. The semen is placed in a small stoppered vial that has been thoroughly cleansed and dried. It may be left at room temperature but it should be kept in a closed container or in a dark place. Under no circumstances must it be exposed to direct sunlight.

If semen is to be held for periods longer than two hours before use low temperatures should be used. The vial containing the semen is filled with liquid paraffin as if for long distance work. It is then wrapped in two thicknesses of paper and set in a refrigerator. Alternatively two rubber thumb stalls can be pulled over the paper cover and the package dropped into a vacuum flask containing water at 37-46°F. Before use, temperature of the semen should be raised gradually, first by exposure to the air and later, if in the field, by placing it against the operator's own body.

More watery types of semen, as from the horse or boar, require different methods of handling. Semen from these animals requires concentration and subsequent dilution, with particular diluents for that of the boar. Further, temperatures for storage of boar semen are slightly higher than those stated, namely, 50-53·6°F., whilst those for concentrated horse semen are slightly lower, 32-37·4°F.

COLLECTION OF SEMEN

Semen can be collected by one or another of several methods. The instrument most commonly used is the artificial vagina. Its size and shape varies, but, fundamentally, it is a rigid or semi-rigid tube made of metal or similar material which has an inside rubber sleeve. One described by workers in Minnesota as suitable for a bull or boar was made from 2¾-inch automobile radiator-hose 18 inches long and a 30 x 3-inch inner tube from a tyre for a lining. An English type, described as the Cambridge pattern artificial vagina, consists of a tin container with rubber sleeve inside to hold hot water, sperm bottle held in place by a brass cap and a felt cover with handle.

The artificial vagina is prepared for use by filling the double walled space, between the outside tube and the inner rubber sleeve, with warm water to bring the temperature of the inner wall to 98-105°F. at the time of use. The amount of pressure on the inner wall may be adjusted by the quantity of water used or by application of air pressure in addition to that of the water. The inside of the rubber tube is greased with a suitable lubricant, and a sperm-catching vessel is attached to the end of the inner tube. The male is allowed to mount a female, another male, or a dummy, but entry is made into the artificial vagina which is held by the attendant in an appropriate position.

A certain amount of training is necessary before males will mount dummies or females not in season. Its object is to secure what is known as a conditioned reflex; that is, to arrange surroundings and use so that when the male finds himself under those particular conditions he will have the reflexes to which he has become accustomed or trained. It is customary to take the male always to the same place and in the same manner. At first he is provided with females that are in season. Subsequently a female not in season can be used or even a dummy. The dummy need not be elaborate. Almost any kind of a frame of suitable size and shape can be prepared. It should have some padding and be covered with a hide of the animal for which it is a substitute. Some males are more vigorous and thus easier to train than others, but that is so for any type of controlled service.

Semen can be collected from bulls by inserting the hand and portion of the forearm into the rectum to massage structures known as the ampullae. A laboratory process adapted for use in sheep (Gunn 1936) is in fairly general use both in Australia and overseas. It is referred to as the method of electrical stimulation. Briefly, an intermittent electric current of particular type is passed through the lower centres of the spinal cord. One terminal is inserted in the rectum; the other is a needle pushed under the skin of the loin. Stimulation of the appropriate nerve centres produces ejaculation, the semen being collected in a test tube or other suitable vessel.

The earliest method used was to collect semen from the vagina of a female served normally. It is still used for sheep, but it has been discarded for cattle, horses and pigs. Great care must be taken that the female is perfectly normal. The vagina of the ewe must be washed free from mucus by repeated flushings with a special saline solution. When all mucus and saline have been removed, coitus is allowed to take place. Afterwards, the semen is drawn up into a special glass tube with a rubber bulb to provide suction. Another method is to insert into the vagina a rubber sperm collector into which entry is made. Still another procedure is that which employs a breeding bag on the penis of the stallion or jack ass. The end result of all processes is to secure semen with as few contaminants as possible.

INSEMINATION

The outstanding advantage of artificial insemination is to permit males to leave more progeny than they could if working under natural conditions. It is achieved, in excess of results from controlled or hand mating, which has the same objective, by making maximal use of each ejaculum.

The following possibilities are disclosed by data published by Dr Walton of Cambridge. The volume of semen obtainable from a single service by a horse is 50-100 c.cs; it can be used for from 8 to 12 mares; that of the bull is about 5 c.cs and can be used for from 10 to 15 cows; the ram emits about 1 c.c. which can be used for from 30 to 40 ewes. Although the boar emits 200 to 250 c.cs, the sperms have low concentration and can be used for only two sows.

Dilution of semen is necessary to secure maximal results; however, an increase over that available from controlled service can be effected merely by fractionating the semen, for example, by using only one-half or one-third or another fraction of the undiluted semen available from an ejaculation. Maximal results from this method are probably from 7 to 10 ewes from a single ejaculum of a ram. Generally, the quantity used is ⅕th to ⅒th of a cubic centimetre. Of course, the quantity used must be associated with the number of sperms in the fluid. For reasons already stated, this must be enormously in excess of the single sperm actually required for fertilization.

Among domestic animals some, particularly mares, receive the ejaculum of the male directly into the uterus, whilst others, notably the cow and ewe, at times other than parturition, have an almost impenetrable canal between the vagina and the uterus. Females of the second kind normally receive the ejaculum into the upper end of the vagina, whence it passes to the canal and so to the uterus and its upper tubes. The size of genital passages also varies as between species. It is necessary, therefore, to have instruments suited to the different animals and to consider their insemination individually. However, one requirement is essential for all. The female must be on heat. Oestrus or heat is the obvious sign of shedding the egg (ovulation) and this must take place before any insemination, whether natural or artificial, can result in conception.

It is possible to induce oestrus and ovulation by injecting substances known as hormones. Multiple ovulation and so abnormally large numbers of progeny at a single birth can also be induced by using these substances. They are secreted by glands within the body and, while circulating in the blood, produce heat physiologically. Many hormonal substances have been analysed and manufactured from chemicals in laboratories. They are available from this source and can be recovered also from glands of animals, in which case they are collected and treated after slaughter. In Russia considerable use has been made of hormonal substances to induce ovulation in large numbers of females during a given period. This simplifies procedure for artificial insemination, and by it females can be caused to conceive "out of season". However, hormonal substances are expensive; possibly in the order of 10s. to 15s. for sufficient to treat one ewe, and the cost would increase proportionally with the size of the dose and of the animal. It is customary, therefore, to examine females and to isolate those that are on heat naturally. It is advisable also to await the final stages of heat rather than to inseminate early in it.

When a mare is in season, the passage from the vagina to the uterus is generally widely opened. Sometimes, however, this passage, which is called the cervical canal, remains rigid, but gentle massage will frequently cause it to relax. This wide-open passage makes the mare the easiest of domestic animals to inseminate artificially. A simple and convenient method described by American workers is to use a half-ounce or a one-ounce gelatine capsule. From 10 to 33 c.cs of semen should be poured into the capsule, and the capsule sealed. It should then be carried into the vagina by hand without delay and placed well forward in the cervical passage. The hand and arm of the operator should be clean and well lubricated.

The method has given place to the more frequent one of using a vaginal speculum and an inseminating syringe. The speculum is a special instrument which can be passed into the vagina in a closed state, when it is relatively small. It can be expanded when in position. It discloses the cervix and provides a tube down which the inseminating syringe is passed. The syringe is first rinsed with and then filled with seminal fluid from which all bubbles are removed. The syringe is carefully inserted through the opening of the passage into the uterus, through the cervix, and the semen is then slowly injected into the uterus. From 10 to 30 c.cs of semen should be injected, the amount depending on the number of mares to be inseminated and the quantity of semen available.

If possible, mares should be inseminated near the end of the heat period, since ovulation occurs about 24 hours before heat ends. The heat period normally is long, ranging from 3 to 7 days and varying from 1 to 15 days or more. It is often difficult, therefore, to tell just when heat begins, and it is sometimes desirable to inseminate twice during the period. Such practice is especially desirable in females which have extra long heat periods and which are difficult to get settled. Breeding the third, fifth, and seventh days, if the mare is still in heat, is considered good practice as the chances of inseminating at the proper time are greatly increased.

Cows to be inseminated should be at the proper stage of oestrus, that is, toward its end rather than at its beginning. Oestrus or heat in these animals usually lasts from 12 to 18 hours and recurs about every 20th or 21st day. Ovulation, that is, the shedding of the egg or eggs, can be expected to take place approximately from 20 to 40 hours after the onset of heat. The optimal time to breed in a heat period depends upon its length and whether one or two services (or inseminations) are to be given. If only one is intended then it should be delayed until from 12 to 20 hours after the onset of oestrus. If two services are planned they could be spaced with advantage, one to be shortly after onset of heat with the other to be from 12 to 20 hours after onset. In many instances onset of oestrus is not observed and no such times can be definitely fixed. The general rule, which covers practical conditions, is that services or inseminations are most likely to be effective if given late in oestrus. If a cow is served soon after she is discovered to be on heat, and she does not remain on heat for over twelve hours, then, other conditions being favourable, the service should be effective. However, if the cow were to exhibit an unduly prolonged period of heat, the service should be repeated.

The cow and the ewe have relatively long, tortuous canals between the vagina and the uterus which generally preclude insemination directly into that organ. All that can be effected is to deposit the semen just a short distance inside the opening of the cervical canal. In the cow, the distance should be in the order of half or three-quarters of an inch, but even this short distance may present difficulties. They arise from backward, finger-like projections which occur on the wall of the canal.

Cows to be inseminated should be secured in a bail so that their rumps are toward the light. This frequently provides insufficient illumination and the operator is then forced to use a headlamp similar to that used frequently by medical men when examining human eyes or throats. A speculum is also necessary. The use of these two instruments makes the cervix, that is, the opening of the cervical canal, visible at the end of the speculum. A syringe, designed as suitable for cattle, is filled with the desired quantity of semen. With the cervix in view the point of the nozzle is gently inserted a short distance (about half an inch) into the passage and 0·5-1·0 c.cs of semen (8 to 16 drops) are slowly expelled into the cervix by gentle pressure on the syringe plunger. It is especially important that only such small volumes of semen as from 8 to 25 drops should be used and that they be injected slowly. If volumes in excess of 25 drops are used, inflammation and fever are apt to set in.

In the U.S.A. a new technique has been evolved whereby semen can be carried into the uterus of a cow and into either horn. No speculum is used. One hand of the operator in the rectum seizes the cervical canal and manipulates it to assist passage of a glass tube pushed forward by his other hand. Available evidence does not indicate a greater percentage of conception by this method than by the other.

Insemination of the ewe is similar to the earlier method for the cow, but the quantity of semen is less. Most workers have found that 0·1 c.c., that is, about two drops of semen, is all that should be injected. For the cow and the ewe quantities have reference to the volume of fluid to be used. If the semen had been diluted the volume would still be the same, although if dilution had been with an equal quantity of diluent, the two drops injected would contain only half as many sperms as the same volume of undiluted semen.

Insemination of ewes is facilitated if they are restrained in some form of crate. This is elevated say 18 or 20 inches off the ground and mounted on an axle so that it can be swung and tilted to suit the operator's convenience. Many forms of crates have been evolved to speed up the work. Generally, they open out of some type of race. Ewes to be inseminated and known to be on heat are worked up the race and through the crate which swings in and out of the race. For reasons already given, inseminations should be made late in oestrus. In sheep its duration varies and ordinary procedure for detection frequently does not disclose the time of onset. For practical purposes, however, any ewe known to have been on heat early on one day and which is still on heat during the following day, should be inseminated a second time. This fixes an interval of approximately 12 hours. Ovulation can be expected to take place within an hour either way of the ewe ceasing to stand for the ram.

USES OF THE PROCESS

Nothing supernatural has been disclosed with regard to artificial insemination. Improved techniques are relatively new, but the process itself cannot be regarded as having recent origin. Further, it is substantially true that the present state of livestock excellence has been achieved without its use. Then wherein does its current interest lie?

Probably a large part has origin in overestimation of what can be achieved by scientific methods. Radiophysics, aeronautics, and the chemistry of explosives, have progressed by leaps and bounds, and rational procedure has been demonstrated in mass production of armaments generally. These developments have induced an eagerness to absorb new ideas. The danger is that many have newness as their greatest merit. Artificial insemination is not wholly of that nature. It has very definite possibilities, particularly as the end stage of selection by progeny-testing.

At a council meeting of the Royal Agricultural Society in England recently, in answer to a question as to how long the period of essential research must continue before the practice of artificial insemination could be employed for the requirements of Central Europe after the war, one of the members replied that the question could not be satisfactorily answered until there had been time to say whether the progeny resulting from artificial insemination could go on reproducing itself. The reply was over-conservative. There are no reasonable grounds for suspicion that progeny resulting from artificial insemination will be different from those begotten normally. It has been shown that, although the technique economizes in sperms, fundamental biological processes of reproduction are not disturbed by it.

Artificial insemination came into prominence when workers in Soviet Russia used it extensively. Language and other disabilities make understanding of the Russian situation extremely difficult. It appears, however, that, after the revolution, neither the number nor the quality of Russian livestock was adequate. Under communal ideology, the problem of their improvement was undertaken directly by the State. Superior sires were imported on a grand scale to carry out wholesale schemes for grading-up native stock. The idea was that all progeny should be by pure-bred sires. This is made clear by the following extract from a report by a Russian worker named Kacalov.

During the years 1935-1939, 9,924,000 horses, 19,160,000 cows, 48,699,900 sheep, and 8,167,400 pigs were served by improved sires. This is not sufficient, however, and in 1941-2 all breeding stocks will have to be served by pure-bred sires only, in order to improve their quality. The number of pure-bred stallions, bulls, and boars on the State and collective farms had considerably increased in 1939 compared with 1935. At the beginning of 1940, however, grade stallions of the first and second generation still formed 41 per cent of horse sires in the State farms and 49 per cent on the collective farms, while the corresponding percentages for bulls of first and second generations were 39 per cent and 64 per cent, respectively. Moreover, the progeny of such sires mated to breedless dams will improve very slowly.

Another worker, Neciporenko· reported on the State plan for development of animal breeding in 1941: "It had reference to the purchase of livestock to establish new collective farms and to replenish small farms. Stocks were to be improved by crossing with stud sires, the sheep with fine-wool rams. Artificial insemination was to be undertaken for 702,000 mares, 2,186,000 cows and 17,900,000 ewes."

With such numbers and objectives, maximal use had to be made of the relatively few available pure-bred sires, which apparently approximated only 50 per cent of those in use as late as 1940. The situation was ideal for the use of artificial insemination; namely, the need to improve large numbers of livestock with too few superior sires to operate normally. The corollary is that conditions the converse of these would not be ideal, and that is the position in Australia. It is the case also in England, the home of many of our pure-breeds, and, incidentally, the source from which Russia drew many of her pure-bred sires.

The foregoing applies to mass activities; however, there are physiological limits to the number of progeny that a particular sire can beget and there are relatively few outstanding sires in each generation. Probably in any breed, at a particular time, they could be counted on one's fingers. Thousands occur that can change bad to good, but very few that can uniformly produce more excellent progeny from very good mates. Artificial insemination would find its most rational use under Australian conditions in increasing the number of progeny by such sires. However, there are optimal limits to the number, and extensive use of any sire distributes his faults equally with his virtues.

The critical condition for use of artificial insemination, therefore, is recognition of suitable sires. There is only one test—the kind of progeny a sire leaves. This has direct bearing upon the question of whether artificial insemination would reduce the number of sires in use and so sales of young males. At first sight it would and much of the criticism now being made in England has origin in the fear that extensive use of the process would restrict bull sales. However, if the practice were confined to proven sires, as it should be, then search for the discovery of suitable sires would be likely to result in the use of even more young bulls than at present. Those suitable as donors for artificial inseminations in subsequent generations and to earn the high fees and notoriety to be associated with such appraisals, could only be discovered by examining the progeny of candidates for selection. Probably as many young sires would be used for breeding, perhaps more than under present conditions, but fewer would be used as extensively or for as long a period.

CHAPTER XV

PROGENY-TESTING OF FINE-WOOL SHEEP

PROGENY-TESTING AND RESEARCH

AUSTRALIA has been carefully breeding fine-wool sheep for over ninety years. For longer than that she has relied largely upon the wool clip to finance the country. Under such circumstances, to admit that insufficient fundamental knowledge is available to implement progeny-testing as it should be carried out, is appalling, but honest and true. The process may be visualized for the future but it can only be improvised for the present.

As has been stated earlier, Mendel established over eighty years ago that characteristics in certain garden plants were inherited independently. For many subsequent years there were two schools of thought with regard to inheritance; that based on Mendel's findings, and that put forward strongly by Galton, which held characters of progeny to be blends of those of their parents. The controversy resulted in a victory for the Mendelian conception. It is now accepted that hereditary characteristics of animals are inherited independently and that the fundamental and unalterable causes of hereditary characterization are paired units of inheritance called genes.

An animal, therefore, must be considered to be a mosaic of independently inherited characters, the aggregate of which can be presented for inspection and evaluation as an individual. At a ram show the judge places rams in their order of merit, judged as animals. In doing so he considers a number of characters independently and collectively. He looks the sheep over for undershot or overshot jaws; for straight front legs and truly aligned hocks. He estimates any dip in the back and feels or looks for "grip", loose shoulders, and other points. Having considered characteristics of the skeleton as best he can in full wool, he examines the fleece; he considers it to be more or less dense; shorter or longer in the staple; to have good or bad colour and handle; to show "breeding" by its regular crimp and general attractiveness; and to have an appropriate tip. He then probably examines covering on the points and assesses degree of uniformity within the fleece as judged by estimates of spinning counts and other attributes. These are only some of the things he will consider. A very large number of others will occur to readers. The point being emphasized is that every single one of those attributes probably is, and at least tentatively should be regarded as, an independently inherited characteristic.

The first necessary step for research is, therefore, that component characteristics, in the genetical sense, should be defined as such, once and for all. The definition must have regard particularly to their independent or dependent nature. Common knowledge associates long staple with open tip and short staple with blocky tip. The two extremes are English long-wools and extra dense wrinkled Merinos. This simple observation suggests that selection for a combination of long staple and blocky tip would be difficult because each appears to be associated with an opposite character. However, adequate research may controvert the assumption, and some long-stapled Merinos do not have really open tips. The question therefore arises: "Is the apparent association real?" Adequately controlled experimentation, designed to isolate and investigate such component characteristics, is the only method by which this and many other similar questions can be resolved. A beginning was made by C.S.I.R. about six years ago, but the programme of research will need to be greatly amplified with regard to personnel and money for many answers to be expected within reasonable time.

When the characters are isolated, that is, when the

mosaic that is the sheep is adequately analysed, it will be necessary to determine how they are inherited. An immense amount of work and large numbers of animals would be required to solve the problems completely. For example, Castle, working on long and short hair and spotted and whole colour, two pairs of characters in rabbits, required 1233 animals to finalize the investigation. For immediate practical purposes, however, finality, although desirable, is not necessary; but knowledge of the method or nature of inheritance is essential. For example, the C.S.I.R. *Journal* of August 1942 published information which disclosed shortened lower jaw to be (i) hereditary, (ii) recessive in character and (iii) apparently dependent upon a small number of factor or gene pairs. Two earlier statements (C.S.I.R. *Journal*, February 1942) reported further investigation of another character, pigmentation in wool. The condition was shown to be (i) hereditary, (ii) recessive in character, (iii) apparently conditioned by a single gene pair, dominance being attributed to the gene that inhibited pigmentation. These two characters fall into the group or class that depends upon small numbers of genes. They can, therefore, be dealt with by established methods appropriate for that class of inheritance.

Interesting and relevant experimental work was undertaken in America (Ritzman, 1920). Rambouillet and Southdown sheep were crossed and then bred on for subsequent generations. Breadth of chest was the character being investigated. Rambouillets are deep and narrow whilst Southdowns are broad and roughly circular, as judged by a cut through the chest from top to bottom. The 50 Southdowns had an average breadth of chest of 25·8 cm. and the 50 Rambouillets had an average chest-breadth of 20·2 cm. Average breadth of 36 first-cross sheep was 20·5 cm. and not one of the 36 had as broad a chest as the broadest Rambouillet. The narrow chest of the Rambouillet was thus dominant over the Southdown type of chest. In the second generation of 41 head, derived from mating first-cross sheep, 31 had narrower chests than the most broad-chested Rambouillet sheep and the breadth of chest of all was near the ordinary chest-breadth of Rambouillet sheep. The remaining 10 head of the second generation were all more broad-chested than the most broad-chested Rambouillet. Their chest-breadth corresponded well with several of the Southdown sheep. The investigation therefore shows that (i) the character is hereditary, (ii) dominance for narrow chest is indicated, and (iii), as is shown in the accompanying table, segregation of the parental types was not clear-cut, as it is, for example, in polledness and hornedness in cattle. In this instance many pairs of genes were necessary to provide the gradient.

TABLE XV

INVESTIGATION OF BREADTH OF CHEST IN RAMBOUILLET AND SOUTHDOWN SHEEP

(RITZMAN'S DATA)

Breadth of chest (centimetres)	*Rambouillet*	*Southdown*	*First cross*	*Second cross*
16				
17	1			
18	1		1	
19	3		1	1
20	13		9	3
21	19		9	6
22	5		9	2
23	6	4	7	11
24	2	1		8
25		7		3
26		8		3
27		17		3
28		6		1
29		7		

Here, then, is a second kind of inheritance. It depends upon many pairs of genes, that is, it is a polygenic character which shows dominance. It will therefore present the problem of the heterozygous individual, the animal that will look narrow-chested but will have broad-chested progeny from ewes of that nature, like the polled bull that leaves horned calves when mated with horned cows. But, because dominance has been established for narrow chest, the character could be extended throughout a flock simply by selecting for it. The progeny test stated is a specialized one and is not that used for characters controlled by simply additive genes although both traits would be classified as polygenic characters.

It is critical, therefore, that the nature of inheritance of independent characters should be discovered. Methods by which they can be controlled vary and depend upon such knowledge. The investigations would take time, but they are necessary before classifications can be made and appropriate action determined.

If the two requirements—definition of component characters and of the nature of their inheritance—were satisfied, it would also be necessary to learn the degree of variation common among progeny of rams being tested for particular characters. This is critical to establish the number of progeny necessary for the test. It is all a matter of sampling. The idea behind progeny-testing is that, by drawing samples of the breeding worth or genotype of the animal under test, a reliable estimate of its real worth can be formed. The number can only be authoritatively stated when the customary degree of variation has been determined. It would depend upon the coefficient of variability, which is the standard deviation expressed as a percentage of the mean. The subject is unavoidably complex; but sufficient has been presented to show the nature of additional experimental work necessary before even the requisite number of progeny can be stated.

There is still another outstanding requirement of knowledge. Every instance in which progeny-testing has been successfully carried out, and there are many, has had reference to a measurable character; for example, egg production per annum or egg size, and yield of milk or butterfat by dairy cattle. Exact measurements lend themselves to mathematical treatment such as that implied in the foregoing paragraph.

If the reader will think back upon recognizable characters of fine-wool sheep, even the relatively few mentioned earlier, he will realize how much experimental work is necessary before each character is measurable. Up to date, fleece-weight, yield, fibre thickness, and so uniformity of fibres, density, that is, the number of fibres per unit of skin surface, and crimp are all capable of measurement. Other fleece characters are under investigation, but as yet the number of known techniques is insufficient to measure all characteristics. It is true that many can be assessed and, as will be shown later, assessments provide very useful approximations; but we are now discussing the requirements of progeny-testing as visualized for use by subsequent generations of breeders. Further, many of the methods now used for measuring fleece characters are too slow for immediate practical purposes. They are suitable for laboratory work, but require speeding-up considerably to be satisfactory for general use.

Before progeny-testing could be used as it should in sheep breeding, all the foregoing points require investigation. The amount and nature of research necessary is obvious, but the hard road is the only sure road. Hazards of breeding must be controlled if wool growing is to be reduced to a level where competition from synthetic fibres can be met. Production costs and thus sales prices are critical. Breeding hazards involve loss in money and thus increased costs. They are inseparable from breeding operations under the control of chance. It gives rise to normal distributions that require to be upset every year by classing and culling. These processes leave the general average and those on the "good" end of the curve, its "tail" end, provide the culls. They are an overhead item that can and should be controlled. For example, if the retained portion be worth £1 per head when culls are worth 10s., then 10s. is lost on every cull.

Costs of wool production depend largely on conditions considered by many of the community to be a necessary part of our social structure. Thus wages and commodity prices could be reduced only by an extensive overhaul of that structure. Heavier clips could, and probably would.

be produced by better feeding, but provision of supplementary fodder is costly. The cheapest feed is that found by the sheep from native herbage in the paddocks. Low costs of production based on that food and on the wages of the time, more than any other factor or combination of factors, placed Australian wool firmly on the English market. They are still essential for its maintenance in that position. Control of overhead by breeding processes likely to reduce culling or to increase levels of fertility and production, and by disease control to restrict losses due to minute organisms or insects, are the two cheapest methods available for probable improvement. Both require exact knowledge before they can be satisfactorily put into practice.

For the method under review there are four points on which further knowledge is required. They are (i) definition of component characters and their associations, (ii) methods of inheritance controlling the characters, (iii) techniques for measuring characters of economic worth, and (iv) degree of variability commonly occurring among the offspring of rams being progeny-tested for the characters.

When sheep-breeders can be given exact knowledge with regard to these four, they will be able to operate with a degree of certainty now impossible. Therein lies the advantage to be gained from such research work as would be necessary. It is not a question of condemning or upsetting existing procedure for stud breeding, but of passing to breeders new tools with which they may be able to work more efficiently. If further steps are necessary before the tools can be manipulated satisfactorily they must also be taken, but only advantages will accrue from exact knowledge applied to reduce production costs.

Reference has been made in Chapter XII to sheep-breeding experiments being undertaken by the Bureau of Animal Industry in the United States. Much of this work is designed to provide satisfactory measurements of fleece characters. The complexity of the problem is indicated by recent results of those researches. Briefly, work to define the number of fibres per unit of skin surface has shown that three areas must be measured on each sheep to estimate adequately variation over the fleece-growing area. Rump, hip, and belly were selected as being representative of the variation. Clean scoured yield also requires to be determined from a composite of three areas—back, side, and shoulder. However, of all single samples, that from the shoulder area is most closely related with yield of the whole fleece, the correlation coefficient being r=0·75 where r=1·0 would be maximal. But these observations were made on fleeces with, by Australian standards, the low yield of 35·4 per cent. They probably are not applicable to Australian wools or to conditions of animal breeding.

Fineness of fibre requires to be estimated from three areas—shoulder, side, and thigh. Further, for a sample to represent a fleece adequately, 450 fibres must be measured, 50 from each sampled area, the diameter of each of the 50 fibres being measured at three levels—high, middle, and low.

These are only some of the problems of measurement. It is equally important to determine to what extent a character is influenced by heredity. Work by Bureau researchers shows this to be low. All of these attributes require to be determined for Australian conditions, but the American findings are of general interest. Results were that differences in staple length were 36 per cent heritable. The proportion in other characters was clean scoured yield, 38 per cent; greasy fleece-weight, 28 per cent; body-weight, 40 per cent; body type score, 12 per cent; extent of face covering, 32 per cent; neck folds, 26 per cent; and body folds, 32 per cent.

Older breeders may think these requirements are fanciful. But Lush (1940), discussing such matters, wrote:

> A dependable estimate of the heritability of a characteristic is perhaps the most important consideration in deciding which of several possible breeding plans is likely to be most effective. If the breeding plan is merely that of mass selection of the most desirable individuals to be parents, the progress per generation

which one should expect is only the heritability fraction (in the narrowest sense) of the amount which is actually reached for when selecting the parents. If the desired characteristics are highly hereditary the best method will be mass selection with little real use for pedigree, relatives, progeny tests or inbreeding. If heritability is low, but there is not much epistatic variance either, considerable use of pedigree and much use of progeny tests and selection on a family basis would be a better plan but there would be little use for any more inbreeding than is needed to make families distinct or to make full use of the progeny test. On the other hand, if there is little additive variance but much epistatic variance, this calls for a considerable use of inbreeding to create lines distinct from each other, with subsequent interline selection and crossing and the creation of future lines from the results of the more favourable of these crosses.

When choosing an efficient breeding system, a usably accurate rough estimate of heritability of the important characteristics is the first thing a breeder needs to know after he has decided upon his goal.

APPLICATION OF THE PROCESS

The foregoing indicates the minimal amount of research necessary before progeny-testing could be used with the degree of exactness justified by its importance. That, however, is an ideal state, and great benefit can be derived simply by using three first principles.

They are: (i) Selection based on the genotype of an animal is more logical than selection based on its phenotype. (ii) The genotype of an animal can be assessed by the average merit of an adequate number of its unselected progeny. (iii) The process is most applicable to characters that are weakly controlled by heredity and markedly influenced by environmental conditions.

Until component genetical characters and their associations are known, it should be assumed that attributes now considered by sheep-classers are characteristics in the strict genetical sense. The nature of their inheritance must be surmised from observations and general knowledge. Broadly, when a character not present in parents appears in their progeny, it can be assumed to be a recessive condition, for example, parrot-mouth.

Conversely, a character of one parent that smothers or

obscures a character of the other parent shows dominance, for example, polledness is dominant over hornedness in cattle. When a character is definite and clear-cut and separates out, or segregates, among relatively small numbers of progeny, then probably it depends upon a small number of gene pairs, for example, the condition hornedness which occurs among polled cattle that are heterozygous for the condition.

Such characters are not best suited for applications of progeny-testing. But, when environmental causes strongly affect the character; when dominance and recessiveness are not apparent; when segregation is ill-defined and gradation takes place from more to less with regard to characterization, then use of the method of progeny-testing is indicated. These occurrences are typical of polygenic inheritance controlled by genes which act additively.

Because it is much simpler to select by few than by many criteria, characteristics should be classified as more or less important. Every sheep-breeder will have his own ideas with regard to fundamental requirements. Some day these may be determined, classified, and stated with regard to competitive attributes of artificial fibres. Until that is done, however, reliance must be placed upon personal opinions of competent classers.

When culling based on them is carried out, all sheep in the flock will be similar in that they comply with those requirements. But the similarity will not be complete. Thus, if an owner were to cull all sheep with less than three inches of staple, those retained would vary from three to possibly five inches in staple length. However, the basic culling would have particular reference to faults that the owner would not condone under any circumstances. He will then appraise his flock with regard to its remaining attributes. By and large it may be short-stapled, too wrinkled, too fine, or too strong, and so on. These are characters for which selection is to be practised to "build them up" in that particular flock. They would not be the same in all flocks, but appraisal of the progeny of any ram or group of rams should be made with regard to them in every instance.

Such appraisals of general weaknesses in flocks largely influence selection and purchase of new rams at, for example, the Sydney Ram Sales. Such rams should be valued by the extent to which one or other of them begets progeny better, in respect to the generally appraised shortcomings of the flock, than the flock ewes with which the rams are mated. Fundamentally, that appraisal of results is progeny-testing, and methods by which it is made are only incidental to it.

The question arises as to whether attributes should be considered separately or regarded in the aggregate. Fortunately work has been done in the United States on this problem.

Hazel and Lush (1942) classified possible methods of selection as of three kinds. The first was called the "tandem method". It is to select for one character at a time until that is improved; then for a second trait; later for a third, and so on, until finally each has been improved to the desired level. The "total score method" is to select for all characters simultaneously, by using some index of net merit constructed by adding into one figure the credits and penalties given each animal according to the degree of its superiority or inferiority in each character. The result is a score of points. In the method of "independent culling levels" a certain level of merit is arbitrarily established for each trait, and individuals below that level are discarded regardless of their superiority or inferiority in other respects.

The investigation showed that selection on total score or index of net desirability is much more efficient than selection for one character at a time. Selection for several traits, by using independent culling levels for each, was more efficient than tandem selection; that is, for one character at a time. Selection on independent culling levels was less efficient than selection on total score, but in some cases it permitted earlier selection without waiting until all characters reached maturity.

It will be recognized that the method of culling for fundamental requirements recommended in the earlier paragraph is that described by Hazel and Lush as the method of independent culling levels. It is evident also that the findings of these workers indicate that point scores, or what they call "the total score method", should be applied to remaining characters.

This introduces desirability of measuring the characters but it has been stated that insufficient basic knowledge precludes adequate measurement. It is necessary therefore to assess merit in these respects. It is advisable but not necessary that assessment should be numerical. That is, instead of saying that a sheep has "excellent" staple length, it should be given 10 points for the character. In descending scale, the "very good" would be 9 points; the "good", 8 points; the "very fair", 7 points; the "fair", 6 points; the "average", 5 points; and so on. The total point score for each sheep is the aggregate of points given to it for every character under consideration. The score of the ram is the total of scores of his progeny. They could be expressed as a percentage of the maximal score to smooth out differences in progeny numbers.

It is obvious that thoroughness in the process of scoring and testing will be associated closely with the value of the sheep. For example, in studs, procedure could well be optimal. Each lot of progeny by each sire could be point scored in the wool when two-tooths. Any previous culling should be noted, because it is most essential that the sires be judged by the average merit of all their progeny and not by a few selected tops. It would be advantageous, too, even if it involved a special shearing of the test groups, to have every fleece weighed and appraised by a competent person so that its commercial value could be determined. After shearing, the progeny groups could be point scored with regard to make and shape, particular reference being made to skeletal faults such as hollow back, knock-knees, and the rest. The score of each sire under test would then be the average scores of his progeny in all three of these examinations.

The procedure might be too complicated for general application. But, provided the three fundamental principles already stated are held and complied with, detail of procedure can be modified to suit conditions of production. Further, it might and probably would be advantageous for studs to have a flock property on which to test their "reserve rams". Any even line of ewes is adequate for the purpose, and to use flock ewes for the tests would conserve the breeding usefulness of highly selected stud ewes. Their matings would then be restricted to top sires.

A less comprehensive test could be arranged in flock sheep after the following manner. The ewes are a run of the flock that has been appraised with regard to the characters in question. That appraisal could be point scored as, for example, 90 per cent for staple length, 60 per cent for plain body, and so on; 100 per cent in each instance being the classer's ideal for his particular requirement. Any experienced judge of sheep could form such an appraisal. Further, it is made, consciously or unconsciously, every day in the week when buyers look over lines of sheep. It is immaterial whether the numerical equivalent of the appraisal is stated, provided only that the appraiser has a fixed standard, that he judges all sheep by it and does not judge them with regard to the average of those he sees before him. For example, a buyer could go to a big sheep sale and point out the best line of ewes offering. He would judge them to be so having regard to the general average of those yarded. But he might look over the yarding and consider that none of those offering were what he wanted. That judgment would be with regard to an ideal. It is only a small step for him to say that the best in the yard were only 75 per cent compared with his ideal, which is 100 per cent.

The rams to be tested are now raddled on the brisket and each is securely and separately yarded with a cut of the ewe flock on say three or four successive nights. Alternatively the rams and ewes could be depastured as small flocks, one ram with perhaps one hundred ewes in as many small paddocks as there are rams to be tested. Local facilities will condition the method. The tupped ewes, discovered by raddle marks, are indelibly marked on appropriate positions with paint, or some such substance, to distinguish them as mates of particular rams. When each ram has tupped about one hundred ewes or any convenient number depending on the size of the flock, the test ceases and marked ewes are depastured as before. For reasons previously stated, the number of progeny is important. It is known that broadly the greater the number the safer the estimate. Further, it is known that twenty is statistically about on the threshold for "population size" to give significant results. Accordingly, the number of ewes tupped by each ram should not be fewer than about thirty, to allow for untoward events, and it should be as large as can be conveniently arranged, fifty lambs being the objective.

When lambing time approaches the marked ewes are isolated from other ewes—but not necessarily from other test groups—so that their lambs can be appropriated to the particular sires. To do this, it is only necessary to ride the lambing paddock at most once daily with a good dog and appropriate paint or ear-marking pliers. By these or some other convenient method the lambs are permanently marked for identification, not as individuals but as members of a sire progeny group. They are to be run later with the general flock so that environmental conditions of the test are those which characterize the property.

At lamb marking it may be possible to observe that the get of a particular ram includes lambs to be culled immediately for outstanding defects, for example, colour, undershot or overshot jaw, and so on. The percentage of lambs begotten by the individual rams should also be noted. Such observations could be made before the ewes and lambs are "boxed" with the general flock.

Other convenient times will occur when all the lambs are in the yards, for example, weaning, lamb shearing, or as two-tooths for classing. They could then be drafted off according to their permanent identification marks. At one or other of these times the test lots could be appraised and their sires valued according to the general excellence of their progeny. The appraisal can be refined to degrees absurd for practical application, or it can be as common-sense as any ordinary commercial judgment. To be worth while it should classify each group of young sheep by its average merit. If the appraiser has sufficient confidence in his judgment to give point scores to the particular pens, that is all to the good. Results from each ram can then be readily compared with the score given to the ewes with which it was mated. Further, the scores for the progeny of each ram can be compared at leisure.

If any sheep-breeder carries the process suggested for progeny-testing as far as having identifiable progeny groups he will find much to interest him and many ways of comparing the groups. Last shearing a number of cellophane envelopes were secured. A twelve-year-old boy was commissioned to stand on the board at one pen with a good-natured shearer and take samples from fleeces during shearing, after or at the third blow on the "whipping side". This inconvenienced nobody. The boy put each sample into an envelope and with it a piece of paper marked, for example, "spot on the head", which identified the sheep as being one of a particular group. Later all the envelopes were collected and classified according to the marks for sires being tested. The staples were tidied up and replaced in the transparent envelopes. No measurements were taken, but the samples were set out in progeny groups and examined through the envelopes to discover differences within and between the various sire lines. To state that the examination was highly informative would belittle its nature. Had fleece-weights and weights of belly wool been taken, or had the fleeces been examined separately by the classer, the data would have been still more informative. The point to be emphasized is that, under present conditions, progeny-testing is very like swimming. The best way to learn how to do it is to jump in. When that step is taken ingenuity and interest in the process will cause sheep breeders to evolve methods of evaluation suited to their own conditions of work. It is hoped that organized research will permit future generations of breeders to be given more specific advice.

CHAPTER XVI

CONCLUSION: CURRENT RESEARCH

It has been a long journey from the Norman manor-house to the Sydney ram sales, but we have arrived. The story of improvement of livestock by breeding processes discloses that a benign association must be maintained between the character and requirements of animals and their environmental conditions. These, whether they be diseases and parasites, levels of available nutriment, high or low atmospheric temperatures or any other, provide a frame within which the animal-breeding craftsman must confine his picture. Adjustments of this nature are essential commitments of animal husbandry in its broadest sense. This statement of principles and methods has reference only to animal breeding as one part of the general problem. It is an essential aspect, but it is not more critical for improvement of livestock than others that have been considered more briefly.

During the period under review it is evident that breeding of improved animals has been guided by practice and knowledge. Of these, practice has been more important. To state their relation metaphorically, improvement and progress can be represented as two individuals who, desiring to move forward, harnessed the horse Practice to a buggy. Practice was staunch and reliable. Although he became bogged on many occasions and seemed unable to avoid rough going, the journey continued. A considerable distance was covered, but Improvement thought they could go faster if they bought another horse. Progress agreed, and eventually they acquired one called Knowledge. He was by Science out of a mare said by some to be Inexperience. In many respects the new horse was different from Practice. He was better looking, but headstrong and more than a little flighty.

The pair did not pull together too well at the start. Knowledge hung back from the collar. At other times he went up with a bound and pulled wide off the pole. Practice held to the track. After the two horses had been working together for some time, both Progress and Improvement agreed they were travelling faster. Practice was still as dependable as ever, but Knowledge pulled him aside from bogs and rough going. Left to themselves, the pair avoided many difficult places. It was noticed that although Practice pulled his weight and more, Knowledge was the pathfinder. This was most evident as the journey lengthened and the going became more uphill in its late stages. However, they could not break Knowledge of one habit. When the horses were turned out at night, Practice stayed around the camp and was easily caught in the morning. Knowledge liked to wander and frequently had to be hobbled to restrain him.

The metaphor describes the present position of animal breeding. Practice and knowledge must each make its contribution. The first is relatively static. However, although knowledge may disclose new possibilities, these must be accepted and applied in practice. The rate of improvement and progress is, in fact, dependent upon the closest association of practice and knowledge. The purpose of this statement has been to facilitate that association. Its conclusion therefore is an opportune place to report experimental work in Australia that will, it is hoped, provide further knowledge.

CURRENT RESEARCH

If the volume and nature of research work in Australia indicate the relative importance to animals of hereditary or environmental causes, then the latter are overwhelmingly predominant. An alternative possibility is that animal breeders require assistance only with regard to problems having environmental origin, that is, those with reference to food and disease. Be that as it may, a survey of research problems throughout Australia discloses nothing comparable, in number or range, with those of other countries that investigate genetics in animals.

Thus, although nearly all State Governments own herds either of beef or dairy breeds of cattle, none uses methods different from those typical of normal procedure in privately owned stud herds. The only experimental approach is to evolve feeding standards.

Every State has fostered herd-testing of dairy cows by one method or another, but only in Victoria is an effort made to emphasize the importance of progeny-tested bulls. In that State propaganda advocating bull indices may be regarded as experimental. It has several aspects. The first problem is sociological: Will dairy-farmers adopt the method and forsake those long established in the breeding of pure breeds of cattle? These are selected records of performance, predominance of pedigree as such, and standard show ring evaluations of type. The second is: If the method of "index breeding" became general for dairy cattle would it, in fact, increase production over State-wide areas? There can be no reasonable doubt the method will achieve its object in herds where food adequate for high production is available. It is problematical, however, whether nutriment of this order is generally available, and equally so how far the level of nutriment in many areas can be improved within the economic limitations of the dairy industry. However, these problems are beyond the range of this discussion.

The third question is one of animal breeding. It has reference to the possible independent inheritance of gene groups controlling high production of milk and butter-fat within dairy cattle populations generally, as distinct from "pure-breds" isolated by registration. The possibility that these gene groups may not be closely associated with any specific "breed points" is great. It was emphasized when reviewing the U.S.A. Bureau of Dairy Industry proven-sire projects (p. 154). If the work of the Victorian Agricultural Department were to demonstrate effectively that these gene groups could be concentrated upon and isolated as well from grade cattle as from pure dairy breeds, repercussions upon the latter would be severe. They could be avoided only by adopting the method within registered groups. This would require extensive modification of existing criteria for their excellence.

Lactation curves have been considered as possibly indicating not only levels of nutriment available for milk production throughout the year but also the hereditary capabilities of animals. Production of milk or total butterfat is recorded, usually at monthly intervals, but sometimes more frequently. These data are plotted against the time intervals and the points so fixed are joined. This provides the curve. Its shape may vary, but under normal conditions it ascends rapidly after the cow calves, remains reasonably flat for a period and descends more or less steeply as she approaches her next calving and is dried off. The Sydney University McGarvie Smith Animal Husbandry Farm is investigating the possibilities of these curves for analysis of dairy production.

Among beef cattle the only genetical problem being investigated is one undertaken by the Council for Scientific and Industrial Research. It is equally ecological in character and has reference to the relative suitability of British-bred cattle and of their crosses with Zebus for the tropical north of Australia. The investigation has been reported and discussed elsewhere (Kelley, 1943). Broadly, it indicates that considerable advantages are probable and that the only apparent, but not insurmountable, disadvantage is the excitable disposition of the crossbred.

Research work in Australia with regard to pigs is stereotyped. Except in so far as efficacy of methods adopted by the various State Agricultural Departments to improve production is itself a research problem in animal breeding, no other of magnitude is being undertaken. Emphasis is placed upon the selection among existing types of those suitable for particular markets and kinds of available food. This is supported by extension services designed to improve management of animals. There is, however, no organized research specifically genetical in character, such as was reviewed earlier for the Bureau of Animal Industry in the United States. Further, except for sporadic litter-testing groups, there is relatively little concentration, for propaganda purposes, upon genotypic selection.

Mass selection, as adopted for poultry, is an equally problematical method of uplift. The procedure widely advocated and used is that of single testing and trap nesting for hens. High producers are thus identified and mated with sons of high producers. Pens of birds with origins of this nature are then tested at recognized centres. Results from such "egg-laying competitions" are widely disseminated in the hope that producers will use the progeny of successful birds to improve the production of their own hens. Results of the method suggest that what may be regarded as a research problem having reference to its efficacy can almost be regarded as having been solved for the negative. For example, although an individual bird has produced 338 eggs in 12 months, the real criterion of improvement, that is, the average production of all birds under test over 11 consecutive years, has shown a downward tendency. The figures for the averages of production by all birds under test in consecutive years for a particular trial in one State are as follows: 202, 198, 199, 187, 193, 195, 190, 185, 186, 185, and 188. These should be compared with results from genotypic selection described earlier as the Sib or Family method (p. 150). Nevertheless, however conclusive the comparison may appear, continued advocacy of mass selection suggests that its efficacy is not yet regarded as having been determined, and the research problem remains.

It is not until sheep are considered that anything approaching a series of planned genetical investigations is discovered. However, for sheep, as for the other animals, feeding as distinct from breeding researches are more frequent.

Genetical investigations of sheep can be classified broadly as having reference to their physiology of reproduction, to breed differences associated with specific environmental conditions, and to planned synthesis or analysis.

Investigations of reproduction in sheep have been undertaken at several centres. Gunn (1943), of Sydney University, has shown that many causes may affect the fertilizing power of semen from rams. Probably the most important contribution of this work has been to establish that atmospheric temperatures over a large part of Australia are sufficiently high to affect the semen of rams adversely. Many that produced normal semen in the cooler months produced only degenerated spermatozoa in the hotter. The advisability of mating in the cooler months is obvious.

Work by the Council for Scientific and Industrial Research has been restricted to ewes. It first investigated probable causes of reduced fertility by contrasting succeeding phenomena of reproduction in highly fertile Dorset Horn ewes, normally fertile Merino ewes, and a group of lowly fertile Merino ewes. Results of this investigation were reported by Kelley in 1937. The seventh, eighth, and ninth conclusions stated in the summary of that report disclose probable causes for inferior lambing percentages from hand service as compared with those from paddock matings.

Ewes extrude their ova in the late stages of any occurrence of heat. Spermatozoa take about five hours to reach the site of fertilization and have a relatively short life in the genitalia of ewes. In paddock matings ewes are tupped several times during their period of heat. Thus fresh spermatozoa are provided to seek the ova in relays. In hand service only one batch is available. If these are introduced close to the onset of a prolonged period of heat, it is highly probable that by the time the ovum is shed the spermatozoa are too devitalized to fertilize it. Hence the returns to services by normal rams and the advisability of remating if the ewe under observation is still on heat on the day following her detection and first service.

The investigation found three probable causes for reduced fertility in the ewes under observation. They were (i) absent or irregular oestrus, (ii) some pathological condition which precluded conception, or (iii) early abortion or resorption of the foetus.

The second and third of the probable causes may possibly, in some instances, be due to endocrine deficiencies and be genetical in origin. But other kinds of pathological conditions also may be responsible. Periodicity of oestrus or heat, however, has been disclosed by further investigation to be closely associated with the heredity of the sheep. For all normal ewes, including Merinos, the calendar year is divided into breeding and non-breeding seasons. All ewes can be expected to come on heat in the autumn and winter and to have a period of sexual inactivity during the late spring and early summer. This is the fundamental rhythm of reproduction. However, it is influenced largely by the heredity of the animals. Merinos have a relatively long breeding season and a short non-breeding season. British breeds and, among those observed, particularly Border Leicesters, have the converse. Crosses between these two have breeding seasons of intermediate length, and Dorset Horn ewes come on heat as early as many strains of Merinos.

These are broad breed differences, but even within the same breed individual ewes behave differently. Some come on heat earlier and some remain on heat longer than others. Also some ewes both come on heat earlier and remain so later than their breed mates. They thus have longer breeding seasons than are customary for ewes of the same kind. These differences are attributed to the heredity of individual ewes and may be associated with varying levels of pituitary activity.

In England similar differences have been observed. Dr Hammond, when in Australia, stated that he was carrying out an investigation to discover which individuals in a group of ewes were capable of producing lambs earlier than their breed mates. He hoped, by selecting for this character during several generations of breeding, to evolve an early breeding strain.

It is well known that in some parts of Australia, Merino ewes do bear lambs to matings made in October and November, that is, to late spring and early summer matings, although these are months which ordinarily would constitute the non-breeding season. Such ewes could be regarded as either a "late" or an "early" strain. Their isolation could have taken place as follows. Pastoralists first mated at a time that suited them best. This may have been either very late or very early in the breeding season. Some of the ewes were discovered to be on heat and bore lambs to very late matings, for example, in the late spring months. Non-breeders were culled or mated again in the following autumn. This would isolate ewes of "late" strains, but not those of "mid-season" or "early" varieties. By continually mating ewes of the late strains and their progeny in the spring, numbers could be increased and a strain developed which would come on heat and breed after other ewes had ceased to do so. However, individual pastoralists cannot be sure whether their own ewes are all of any particular variety. They are advised, therefore, to test the heredity of their ewes in this respect. This can be done with small pilot flocks, and appropriate mating times can be chosen during which ewes will be receptive to the ram.

These investigations are being deflected from genetical aspects to determine whether the occurrence of green feed affects the rhythm and whether, as an aspect of management, ewes behave differently when permanently or occasionally depastured with rams.

The time of onset of the breeding season was also

investigated by Thompson (1942) at Roseworthy, South Australia. His problem was to secure early lambs from crossbred ewes. He found that the breeding season of Dorset x Merino ewes commenced and reached a maximum earlier than that of South Australian Merinos. Although Border Leicester and Romney Marsh half-breds did not reach their maximum until February-March, the onset amongst the Border Leicester half-breds was earlier than that of the Romney Marsh cross-breds. Both groups by the long-wool rams from Merino ewes were considerably later in their breeding season than Merinos or half-bred Merino x Dorset Horns.

Daley and Eastoe (1943) have reported the average duration of pregnancy of Dorset Horn ewes at Wagga to be 144 days, or four days shorter than was stated earlier. Shier and Underwood (1941) in Western Australia have observed increased ovulation rates and so a greater number of twins as the result of "flushing" ewes under observation. "Flushing" depends on a rapid improvement of the available food just prior to mating. An ideal arrangement would be to have the ewes in strong "store" condition about six or eight weeks before mating time. They would then be fed a much improved ration and be mated when they were doing well. The process does not expedite onset of the breeding season, but because more ova (eggs) are shed, a greater number of multiple births occur.

The New South Wales and Victorian Agricultural Departments have planned a long-term investigation to discover which of the crosses of British and Merino breeds are the best mothers of export lambs. The proposal is to mate Merino ewes on the same property with different breeds of British rams. The ewe progeny then will be divided between the two Departments and, in each State, they will be mated with appropriate rams. The experiment is in its early stages.

Possibilities and hereditary advantages of different British breeds and their crosses with Merinos for the export lamb trade have been and are being investigated by most of the State Agricultural Departments. Results are published from time to time in their respective journals. Broadly, they show that first-cross ewes are the most profitable mothers; that Dorset cross ewes come on heat earlier than other crossbreds and thus are well adapted to bear lambs which must be born early to take advantage of green feed that has a restricted growing period, for example, subterranean clover in a climate with wholly winter rain, such as Western Australia; that Southdown sires produce the best export lambs but their carry-over value is low compared with lambs by long-wool rams; and that Dorset Horn rams are considered to beget, from suitable ewes, lambs which reach profitable market weight and condition in the least time.

Reaction of strains of Merinos, broadly defined as fine-wool and strong-wool, to different levels of nutriment is being investigated by the C.S.I.R. in Adelaide. The investigation is nutritional rather than genetical in character. Nevertheless, it is most informative with regard to instability of the phenotype in wool production.

The experimental animals are paired ewes that have been under observation since weaning. The fine-wools are from New England (New South Wales). The strong-wools are twin pairs of South Australian Merinos. One of each pair of both kinds is fed adequately while the other is fed inadequately. The two rations are referred to as high and low planes. The difference in average fibre diameter of the two lots, that is, strong and fine, on the high plane in 8 microns.* Their spinning counts are approximately 56/58s (28 microns) for the strong-wools and 64/66s (20 microns) for the fine-wools. This probably discloses the extent and nature of their hereditary difference but, when the two lots on the low plane of nutriment are compared, the difference is reduced by 55 per cent (4·4 microns). The average fibre diameter of the strong-wools was 21·5 microns (spinning count about 64s)

* A micron is the unit of microscopic measurement. It is written as the Greek letter μ, like a "u" with a long tail. It is the equivalent of 1/10,000 of a centimetre or of 0·00003937 of an inch.

and that of the fine-wools was 17·9 microns (spinning count about 70s).

Body size and fleece-weight also are markedly affected by the plane of nutriment. Before the second shearing, average body-weights for those on a high plane were 147·4 lb. for the strong-wools and 114·4 lb. for the fine-wools. On the low plane, however, both lots, strong and fine, averaged only 65 lb. The difference in the ration caused those on the low plane to give only about half as much wool as their mates on the high plane. The figures are: strong-wool, high plane 18·4 lb., low plane 8·0 lb.; fine-wool, high plane 13·3 lb., low plane 7·3 lb. There is also a corresponding difference in conformation and degree of wrinkling (development) in the two groups.

The experiment is designed to determine the final effect of early nutrition on the wool of the grown sheep. For the present purpose it is extremely interesting in that it shows the true value of phenotypic selection for characters largely affected by environmental forces. The strong-wool pairs are twins. Some may be from the same ovum and thus be identical; in any case their heredity was derived from samples of the same parental genotypes. When judged by the above criteria of fleece-weight and spinning count, the lot on the high plane would be typical strong-wools, while their twin mates on the low plane would be medium-fine. Further those of the high plane group have grown twice as much wool as their twin mates. These data emphasize again the insecure nature of phenotypic criteria for selection and the interdependence of hereditary and environmental causes.

Wool types for very many years have been determined by trained appraisers. Latterly, however, a move has taken place to supplement appraisals by actual measurement of wool characters. This has become desirable because correlation studies, to determine cause and effect, require accurate quantitative assessments.

Many wool characters are too indefinite to be measured. But density, that is, the number of fibres per unit of skin surface, can be measured indirectly and directly. Other characters now being measured directly at Australian research centres are fibre diameter, which differentiates strong and fine wool and is associated with spinning counts; contour,* that is, the nature of the straight edge of the fibre showing whether its thickness is uniform throughout, like a lead pencil, or pinched off at places by starvation, fever and other causes, when the contour is in peaks and troughs instead of being straight; degree of circularity, that is, whether, on cross section, all the fibres of a staple appear equally circular; and the nature of the difference between the short and long diameters of oval-shaped fibres. In addition, yield, that is quantity of clean scoured wool recoverable from greasy fleeces, is being determined, and endeavours made to define "handle" and other characters now only qualitatively assessed.

Measurement of these characters is being used in a number of researches, many of which have reference to nutritional effects. For example, fibre fineness of wool from the paired sheep on high or low planes of nutrition was determined by direct measurement. This also applies to work at the Institute of Agriculture, University of Western Australia, in which State climatic conditions customarily produce "tender wool" during the summer. This has been shown to be due to an annual protein deficiency. The work has also shown that body-weight falls considerably before wool growth is greatly affected. The sheep maintain this by withdrawing protein principally from their muscles.

At the Gordon Institute of Technology, Geelong, Victoria, an endeavour is being made to measure wool characters of different breeds and crosses and thus to define wool types. The measurements are being used to associate them with manufacturing requirements.

The C.S.I.R. in New South Wales is using similar methods, particularly the determination of curves of fibre

---

* The word "contour" is used by S. G. Barker (1929) and others to denote the contour of cross-section of fibres; here, however, it is used according to the definition now given.

diameters derived from measuring a number of fibres in sample staples of wool. The curves represent the degree of uniformity in a staple, an attribute of wool much desired by manufacturers. Until it is shown that the character of these curves is hereditary, their determination can only be a refinement of phenotypic appraisal. Accordingly, this research is being applied to sire and progeny groups. It is also necessary to show that the characteristic shape of the curve is a constant peculiarity of the individual. This is being investigated by measuring samples from the same site on the same sheep over long periods of time. If the shape of these curves measures a desirable hereditary character, it is necessary to determine how early their shapes become constant. This is being investigated by examining samples taken periodically from lambs at one month and greater ages.

Other breeding investigations are being carried out at the same centre. One of these has reference to the inheritance of wrinkling. Broadly, the plan is to mate, in the parental generation, wrinkly Merinos reciprocally with Border Leicesters as a plain-bodied type. Their progeny are being mated together to determine the extent of wrinkling in subsequent generations. Back-crosses will be arranged. An endeavour is also being made to associate the degree of wrinkling with the number of fibres per unit area of skin surface. These sheep are also providing, in sire progeny groups, wool samples that are used to investigate the inheritance of curves for fibre diameter. They are being studied with regard to their form in the different generations of the crossbreds.

This centre has reported also (Kelley and Shaw, 1942) upon the character pigmented wool. It has shown that when it is defined as large patches or wholly pigmented fleeces, and white wool is the absence of these, then pigmented wool is a simple unit-factor recessive condition. It was suggested that the two, white and pigmented wools, must be associated with the presence or absence of genes of inhibiting nature. However, the pattern of pigmented wool, that is, how it occurs over the fleece and the intensity of the pigmentation, appear to be two separate polygenic characters. It was shown that the occurrence of very small pigmented areas, some of which were minute and carried very few pigmented fibres, increased with age. Also they were more characteristic of some strains of sheep than of others. The hereditary nature of polledness in sheep and of skeletal faults in Merinos is also being investigated at the same centre. Shortened lower jaw as a recessive condition (p. 126) is one of its findings in this respect.

The foregoing review, unfortunately, is important particularly with regard to the omissions it discloses within the restricted programmes of animal breeding research. Whether these, or any large number of them, will be made good during the period of post-war reconstruction remains to be seen. However that may be, mass selection, as widely practised, will carry animal populations thus far but no further. Continued progress is possible only by the application of exact knowledge and genotypic selection. Economic pressure eventually will compel this, and every country depending largely upon animal production, to accept and apply those two principles.

Other countries are aware of these requirements. In addition to extensive use of artificial insemination to grade up her flocks and herds, Russia has many animal breeding laboratories. From heterogeneous ancestry the Soviet Siberian-Rambouillet has been developed since 1926. Rams of the breed weigh 237 lb. live weight and shear up to 30 lb. of wool with 3½ inches of staple length in combing types suitable for fine worsteds. Recent conversation with a member of the staff of the Moscow Institute of Sheep and Goat Breeding confirmed this development. He stated the new Russian type of Merino, of high quality, weighed between 200 and 225 lb., and shore 31 to 36 lb. of 60/64's wool with a yield of 35 to 40 per cent. Progeny-performance has been the mainstay of procedure for its development.

Attention has already been drawn to methods used in Denmark to uplift production levels of pigs and dairy cattle. Similar undertakings have also been discussed for the U.S.A. The following extracts from the 1944 annual report of the Bureau of Animal Industry with respect to research in animal husbandry projects is, therefore, appropriate. In addition to other investigations it reports results of steer-progeny trials after three generations of Shorthorn sires tested in record of performance feeding tests. Steers sired by the third bull, selected by the method in the series, were superior in average daily gain and in efficiency of feed utilization. They also required 22 days less time to reach a slaughter weight of 900 lb. live weight than the steer progeny of the second selected bull and 41 days fewer than progeny of the first bull used in the experiment. Steers by the third proven bull dressed 61·6 per cent whilst those of the earlier bulls were 60·9 per cent and 59·4 per cent respectively. These results show that progress can be made in efficient production of beef by using bulls proved for their ability to transmit such qualities as rapid rate of growth, gain and capacity for economical use of food.

Shorthorn x Hereford steers born in the experimental cross-bred group already mentioned (p. 191), made 70 pounds more feed-lot gain than straight-bred Herefords; and their carcasses were of similar quality. In both carcass grade and gains in the feed-lot, the second-cross steers (Aberdeen Angus x Shorthorn-Hereford) were superior by 26 per cent in gross returns to either the first-cross or the Hereford steers. These observations clearly indicate possible gains from hybrid vigour in producing commercial cattle.

Reports of sheep breeding projects include a statement that considerable improvement has been made in recent years through development of accurate laboratory measurements for fleece characters and their application in the selections and matings of the breeding animals. Weight of wool has been increased 14 per cent; yield has been increased 5 per cent; and staple length, 16 per cent, within four years. One flock being line bred by progeny test for staple length has increased 16 per cent in this respect. The entire Rambouillet ewe flock shows a 28 per cent increase in strictly combing wool types.

Similar achievements were reported from the swine breeding laboratories where a wide range of problems is being investigated. Poultry also received attention. Incidence of Fowl Paralysis had been reduced from 47·2 in a susceptible strain to 11 per cent in one line bred for resistance to the disease. Deterioration rate of egg-white had been reduced by 12 per cent in progeny of a non-deterioration line-bred flock. Shell quality was improved through family selection and, whilst daughters by selected line-bred sires averaged 199 eggs per bird, daughters of selected dams averaged 207 eggs per bird per annum.

Every section of the report disclosed improvement by use of breeding methods based upon record of progeny-performance associated with some degree of inbreeding to consolidate the gains. Australian stock breeders should give serious thought to these findings. The high levels they have attained with some species will be difficult to maintain and low levels of others are uneconomical.

APPENDIX I

WRIGHT'S COEFFICIENTS OF INBREEDING AND RELATIONSHIP

RELEVANT biometrical studies and formulae by Dr Sewall Wright disclose two principal considerations. The first is relationship between animals. This may be between those not mated together but having common ancestry, or between parents.

Measurement of relationship between animals provides evidence of probable similarity. The expectation depends upon the knowledge that closely related animals owe at least part of their heredity to contributions from the same genotypes. When parents are related their offspring are to some extent inbred. Biometrical appraisals of relationship between parents therefore estimate the degree of inbreeding of the progeny.

A second purpose of the formula which assesses degrees of inbreeding is to determine the extent to which homozygosis has been increased by the process. The base line for an increase is the degree of homozygosis which characterized the general population. For example, if all Shorthorns were heterozygous for 200 pairs of genes, then cattle with coefficients of inbreeding equal to 25 per cent could be expected to be homozygous for one-quarter of the number, that is, for 50 pairs of genes, and to be heterozygous for the remaining 150.

On the other hand, if the average Shorthorn were heterozygous for 1000 gene pairs or groups, then a beast with a 25 per cent coefficient of inbreeding would be homozygous for 250 and be heterozygous for 750.

The coefficient of inbreeding thus does not measure an exact state or degree of homozygosis or heterozygosis but only the state of the animal relative to the average of the population from which it came. This will vary from breed to breed and from herd to herd. The coefficients, therefore, although descriptive, are not suitable criteria for other than restricted comparisons.

Practical applications of the assessments are (*a*) appraisal of probable prepotency where this is largely dependent upon homozygosis; (*b*) anticipation of the possible occurrence of defects and faults due to recessive conditions; and (*c*) avoidance of risks from inbreeding by confining matings to those which maintain the lowest possible coefficients of inbreeding. Genes responsible for defects are as surely fixed, if they are present, as those that are advantageous; but they are least likely to become fixed among heterozygous animals.

THE COEFFICIENT OF INBREEDING

Wright's formula for the coefficient of inbreeding is:

$$F_o = \Sigma\left(\tfrac{1}{2}^{ns+nd+1}(1 + F_a)\right)$$

In this formula the symbols are: F, the coefficient of inbreeding; F_o being that of the offspring which is to be determined, and F_a that of the recurrent ancestor if it is inbred. The sign Σ conventionally represents "the sum of". It implies that the formulae inside the brackets should be calculated for every line of descent traceable from the sire to a common ancestor and back again to the dam, and that all the results so obtained are to be added together. The fraction $\frac{1}{2}$ has its face value. The letters *ns* represent the number of generations from the sire to the common ancestor and *nd* represents the number from that ancestor to the dam. The figure 1 has its face value. It is included on mathematical grounds. Figures substituted in the working for *ns* and *nd* and the figure 1 are to be added together. This is indicated by the plus signs. The sum of these three determines the number of times $\frac{1}{2}$ is to be multiplied by itself. The mathematical expression is "raised to the power of –", the figure determined occupying the position of the dash. It is conventionally written next to but slightly above the number that is to be raised, for example, $\frac{1}{2}^2$ means $\frac{1}{2}$ raised to the second power or $\frac{1}{2} \times \frac{1}{2} = \frac{1}{4}$. If the sum of the three symbols *ns*, *nd*, and 1 were 3, then $\frac{1}{2} \times \frac{1}{2} \times \frac{1}{2} = \frac{1}{8}$ would be implied; if it were 4, the product would be $1/16$ and so on.

The formula inside the smaller bracket is only used when the common ancestor, or recurrent parent, is itself inbred. The figure 1 has its face value and F_a, which is the coefficient of inbreeding of the recurrent parent, is to be determined by counting paths as generations and multiplying $\frac{1}{2}$ by this sum plus 1.

Theoretically, a coefficient of inbreeding can be unity. This occurs only when repeated self-fertilization takes place. It therefore is not found among data from animals. Fractions determined by these calculations can be expressed as common fractions, in decimals or as percentages, but all have the same meaning, e.g. $\frac{1}{4}$, 0·25, and 25 per cent are different forms in which a coefficient of inbreeding having identical values can be stated. The second and third are most commonly used.

It will be realized that the critical arithmetical process is the multiplication of $\frac{1}{2}$ by itself as many times as is indicated by the sum of the paths plus one. Calculations of the coefficients are therefore facilitated by the preparation of a reference table which shows results of these multiplications. Such a table is as follows. It can be extended further if this is found to be necessary.

			Per cent
$\frac{1}{2}^1 =$	$\frac{1}{2} =$	$0{\cdot}5 =$	50
$\frac{1}{2}^2 =$	$\frac{1}{4} =$	$0{\cdot}25 =$	25
$\frac{1}{2}^3 =$	$\frac{1}{8} =$	$0{\cdot}125 =$	12·5
$\frac{1}{2}^4 =$	$\frac{1}{16} =$	$0{\cdot}062 =$	6·25
$\frac{1}{2}^5 =$	$\frac{1}{32} =$	$0{\cdot}031 =$	3·125
$\frac{1}{2}^6 =$	$\frac{1}{64} =$	$0{\cdot}0156 =$	1·56
$\frac{1}{2}^7 =$	$\frac{1}{128} =$	$0{\cdot}0078 =$	0·78
$\frac{1}{2}^8 =$	$\frac{1}{256} =$	$0{\cdot}0039 =$	0·39

The first example is deliberately simple and uses only half the formula because the recurrent ancestor is not shown to be inbred.

Example I (Fig. 48)

In the pedigree C is the recurrent parent or common ancestor. Genes that C passes to A would have a chance of being duplicated or paired with others of the same kind passed by C to B. They could thus become homozygous. As a routine procedure, the easiest way to count the paths is first to rewrite the sire and dam of the animal for which the coefficient of inbreeding is being determined. Then write the common ancestor. In the example it is C. Now draw a line from C to the sire and another from C to the dam. Each of these lines represents a generation. There are two of them. This number is added to one making a total of three. Reference to the table shows that $\frac{1}{2}$ multiplied by itself three times, or raised to the third power, is $\frac{1}{8}$. This may be stated also as 0·125 or as 12·5 per cent. It is coefficient of inbreeding of the animal O in the first example.

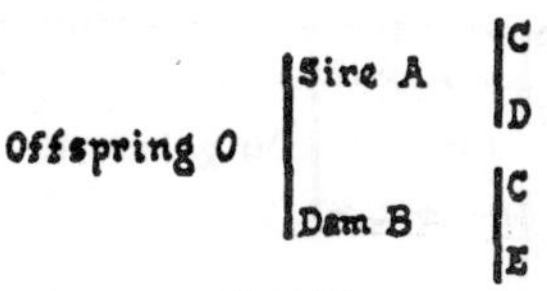

FIG. 48.

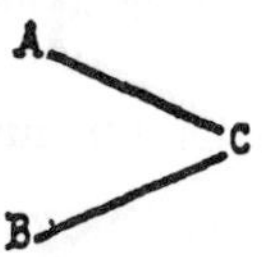

FIG. 49.

Example II (Fig. 50)

Solution: Again write down the sire and dam and identify the common ancestor or ancestors. It will be seen that F and G

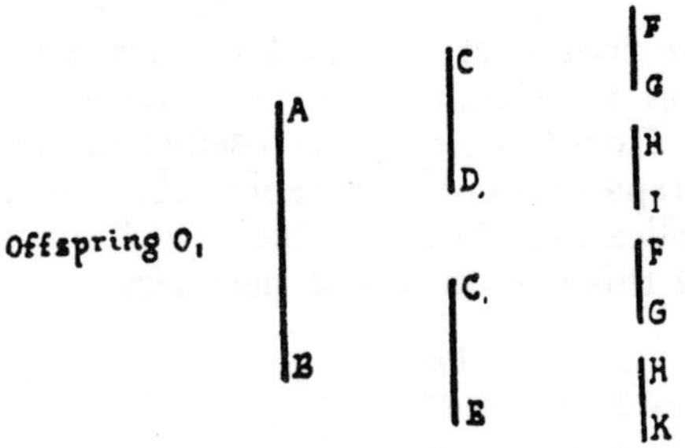

FIG. 50.

appear in the sire's pedigree and also in that of the dam. They are thus common ancestors, but they are not considered as such, because C is taken and his contribution is not only greater but also made possible by what he received from his parents F and G. C is again an ancestor common to both the sire and dam (A and B) and the calculation is as before. These two paths are shown as whole lines.

In the pedigree H is another recurrent parent. It is written down and again paths representing generations to A and B are traced (these are shown as broken lines). From H to D is one generation, from D to A is another, making a total of two. (The little dot on the broken line from H to A represents D.) Now on the dam's side; H to E being one generation makes a sub-total of three, and that from E to B brings it to four. (The broken line from H to B shows these generations and the dot represents E.)

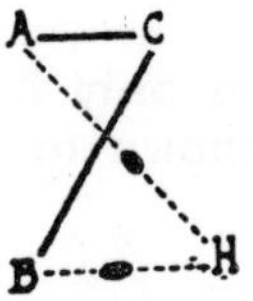

FIG. 51.

The number one is added to the four determined to be the number of paths from the common ancestor to the sire and dam of the animal for which the coefficient is being determined. Summation gives a total of five. The power by which the fraction $\frac{1}{2}$ is to be raised is therefore five. Reference to the table will show that $\frac{1}{2}^5 = \frac{1}{32}$, 0·031 or 3·125 per cent depending upon which system of figuring the operator prefers. Neither of the recurrent ancestors C or H is shown to be inbred, so the data are complete.

There is a result from each of these calculations. The symbol Σ in the formula shows that the two results, determined independently, one for each recurrent parent, must now be added together. For convenience their expressions as percentages are used for this process.

The coefficient of inbreeding of O_1 (i.e. F_{o1}) thus becomes:

	Per cent
Inbreeding due to recurrence of C =	12·5
Inbreeding due to recurrence of H =	3·125
Total inbreeding	15·625

Example III (Fig. 52)

Solution: Proceeding as before, make a diagram and count paths from each of the recurrent ancestors to the sire and dam of O_2.

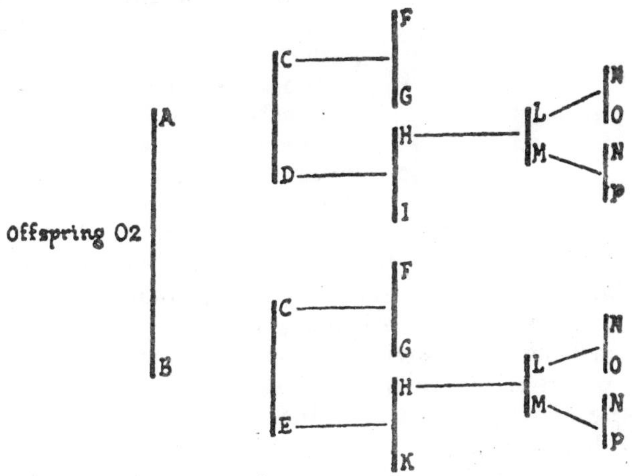

FIG. 52.

It is now shown, however, that H is itself inbred and an appropriate allowance must be made for this. It requires the whole formula as first stated. In addition to the figuring already done in Example II, the coefficient of inbreeding of H must be determined.

There are two paths, one from N to L and one from N to M, which with one added make three. Half must be multiplied by itself three times which is $\frac{1}{8}$. This is the coefficient of inbreeding of H or, as is shown in the formula, of F_a, because H is a recurrent parent which is inbred. The coefficient representing the degree of inbreeding of O_2 due to the recurrence of H thus becomes, by substituting in the formula

$$= \tfrac{1}{2}^5 (1 + \tfrac{1}{2}^3) = \tfrac{1}{32} \times (1 + \tfrac{1}{8}) = \tfrac{1}{32} \times \tfrac{9}{8} = \tfrac{9}{256} = 0{\cdot}035$$

alternate figuring is:

$$= 0{\cdot}031 \times (1 + 0{\cdot}125) = 0{\cdot}031 \times 1{\cdot}125 = 0{\cdot}035$$

As a percentage, $0{\cdot}035 \times 100 = 3{\cdot}5$ per cent.

The coefficient of inbreeding of O_2 in Example III thus becomes:

	Per cent
Inbreeding due to recurrence of C =	12·5
Inbreeding due to recurrence of H =	3·5
Total inbreeding	16·0

Example IV (Fig. 53)

Solution: A is the recurrent parent in the pedigree of O_3. He appears as the sire and a grandsire. There is no generation or path from A as the recurrent parent to A as the sire of O_3. There is one path from A to the dam B. The symbols $ns + nd + 1$ in the formula thus become $O + 1 + 1 = 2$. Half to the second power is $\frac{1}{4}$, 0·25 or 25 per cent which is the inbreeding due to the recurrence of A.

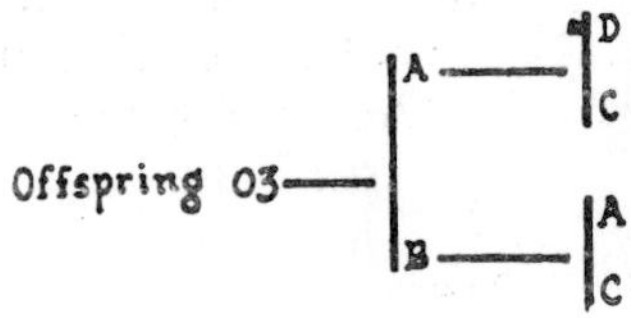

FIG. 53.

C is another recurrent parent. There is one generation from C to A, the sire, and another to B, the dam. These make two. Add one to that number and half has to be raised to the third power which is $\frac{1}{8}$ or 12·5 per cent.

The coefficient of inbreeding of O_3 in Example IV, therefore, is 37·5 per cent, being the sum of the inbreeding due to the recurrence of A = 25 per cent and that due to the recurrence of C = 12·5 per cent.

If these processes are applied to each recurrent ancestor in turn and the results totalled, coefficients of inbreeding shown for animals throughout the text can be checked as further exercises.

Four coefficients of inbreeding have been determined in these examples. They are:

	Per cent
Coefficient of Inbreeding of O =	12·5
,, ,, ,, ,, O_1 =	15·625
,, ,, ,, ,, O_2 =	16·0
,, ,, ,, ,, O_3 =	37·5

Average of the four coefficients of inbreeding is 20·8 per cent. These four animals might comprise a small flock or herd within any breed. If the average reduction of heterozygosis in that breed were 5 per cent, this small herd should be 15·8 per cent more homozygous than the average of the breed. Within the herd O_3 would be 25 per cent more homozygous than O. Similar differences are obvious for the other animals. Broadly, the magnitude of these differences indicates greater or lesser expectation of prepotency due to individual homozygosis in respect to additive genes.

THE COEFFICIENT OF RELATIONSHIP

As stated in the text, Wright's formula for the coefficient of relationship is more formidable than either that for the coefficient of inbreeding or for "percentage of blood" relationships. Its complexity is due to adjustments for inbreeding and for collateral relationships. However, if little or no inbreeding is disclosed by a pedigree, then calculation of the coefficient of relationship is quite straightforward. Broadly, results from its calculation differ from those for "percentage of blood" relationship only when inbreeding is evident. The formula for the coefficient of relationship, which expresses in terms of correlation the probability of two animals, X and Y, having the same genes and so being similar, is:

$$R_{XY} = \frac{\Sigma(\frac{1}{2}^{n + n1}(1 + F_a))}{\sqrt{(1 + F_X)(1 + F_Y)}}$$

The symbols in the formula are, R signifies relationship; X and Y are the two animals being considered; R_{XY} thus means the relationship between the animals X and Y. The numerator of the fraction is similar to the formula for inbreeding except

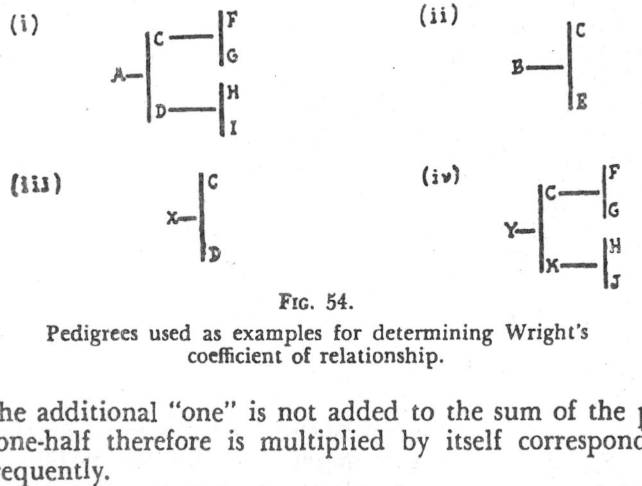

FIG. 54.

Pedigrees used as examples for determining Wright's coefficient of relationship.

that the additional "one" is not added to the sum of the paths. The one-half therefore is multiplied by itself correspondingly less frequently.

The denominator of the fraction is the square root of the product of 1 + the inbreeding of one of the animals (F_X) and 1 + the inbreeding of the other (F_Y). When the animals are not inbred the numerator, and particularly that part in which $\frac{1}{2}$ is raised to the power determined, becomes the coefficient of relationship.

Where there is no inbreeding, the coefficient cannot be higher than 50 per cent. However, in pedigrees where there has been a considerable amount of inbreeding theoretically the relationship can approach 100 per cent. It will be remembered that Wright himself showed that Bates held the Duchess family of Shorthorns with 60 per cent relationship to the bull Favourite during approximately 40 years of breeding.

Fig. 54 presents four simple pedigrees. They will be used as examples to explain the method of calculating the coefficient of relationship between letters which represent animals appearing in the pedigrees.

The relationship of the animals A and B in (i) and (ii) above is $\frac{1}{2}^2$. They are related through C which is one generation back[*] from A (n in the formula $= 1$) and one generation back from B (n_1 in the formula $= 1$). The pedigrees show no relationship between C and D nor between C and E; nor is there any apparent relationship between D and E.

A and B therefore are simple half-brother and sister. Their probable degree of resemblance for characters determined entirely by heredity dependent on additive genes is expressed by a correlation of $\frac{1}{2} \times \frac{1}{2} = \frac{1}{4} = 0{\cdot}25$ or 25 per cent.

Animals A and X in (i) and (iii) above owe their relationship to their common ancestors C and D. They are full brother and sister. There is one generation from C to A and one from C to X. Half is therefore multiplied by itself twice, n and n_1, in the formula, being one in each case. Further, there is no evidence of inbreeding. The result therefore is $\frac{1}{2}^2 = \frac{1}{4} = 0{\cdot}25$. Similarly, there is one generation from D to A and one from D to X, i.e. $\frac{1}{2}^2 = 0{\cdot}25$. The sum of these two fractions $(0{\cdot}25 + 0{\cdot}25)$ is $0{\cdot}50$ or 50 per cent which is the coefficient of relationship between A and X.

Animals A and Y in (i) and (iv) above are related through two common ancestors. Both have C for a sire and both are out of dams which were sired by H. C is one generation from A and one from Y. Their relation because of this is therefore $\frac{1}{2}^2 = 0{\cdot}25$ or 25 per cent. H is two generations back from Y, and the same number from A. Their relationship due to H therefore is $\frac{1}{2}^4$ or $1/16 = 6{\cdot}25$ per cent. The coefficient for the total relationship between A and Y, as disclosed by their pedigrees, is therefore $25 + 6{\cdot}25 = 31{\cdot}25$ per cent.

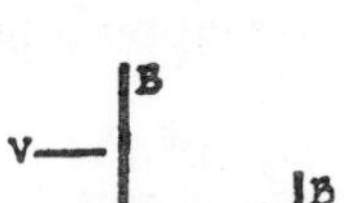

FIG. 55.

Pedigree of V with duplication of B.

Relationship may exist also between an animal and its direct descendants. Thus in Fig. 55, V is shown to have been produced by mating B to his own daughter. The only ancestor that B and V have in common is B himself. The number of generations from B to B is zero (O) i.e. $n = 0$. The number from V to B is one on the sire's side, i.e. $\frac{1}{2}$ or $0{\cdot}50$; and the two on the dam's side, i.e. $\frac{1}{2}^2 = 0{\cdot}25$.

In this case, however, V is inbred so the denominator of the fraction, which is compounded from the formula stated for the

[*] In counting generations for the coefficient of relationship, the paths end at the animals involved. In the coefficient of inbreeding, they end at the sire and dam of the animal in question.

coefficient of inbreeding, must be considered. The degree of inbreeding of V must be calculated by the method explained in the earlier section. V is inbred because B is a recurrent ancestor; *ns* (in the first formula), i.e. paths to the sire of V from B are zero (O); *nd* (in the first formula), i.e. paths from the recurrent ancestor to the dam of V total one. The value of $ns + nd + 1$ (in the first formula) thus becomes $O + 1 + 1 = 2$. Half is raised to this power ($\frac{1}{2}^2$) and the result is $\frac{1}{4} = 0 \cdot 25$. The denominator of the second formula thus becomes, by substituting 0·25 for F_x, $\sqrt{1 + 0 \cdot 25} = \sqrt{1 \cdot 125} = 1 \cdot 06$. The coefficient of relationship of B to V therefore is the sum of the values determined for n and n_1, which arc $(0 \cdot 5 + 0 \cdot 25)$ as the numerator, divided by 1·06 which is the denominator of the formula when substitution of numbers for symbols is completed. The quotient is 0·708. The coefficient of relationship of the two animals B and V therefore is 70·8 per cent.

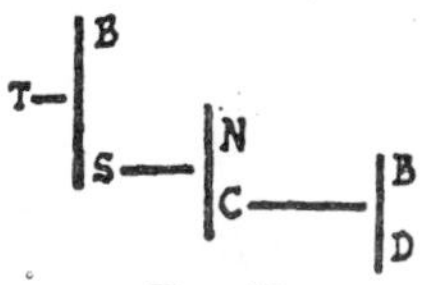

FIG. 56.

Pedigree of T with B occurring twice.

Fig. 56 shows another pedigree similar to that of V because both T and V are line bred to B. However V resulted from mating a sire to his own daughter, whereas T is the offspring by B from his granddaughter.

Analysis of the pedigree of T shows this animal to have a coefficient of inbreeding of 0·125 or 12·5 per cent. The coefficient of relationship of B to T is determined as for V. B to T is

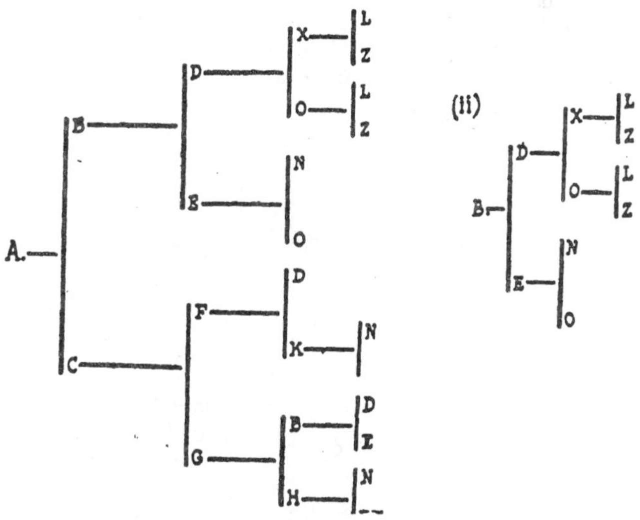

FIG. 57.

Pedigree of demonstrate collateral relationship.

one generation on the sire's side = 50 per cent. B to T is three generations on the dam's side = 12·5 per cent. The denominator of the formula is the square root of $1 + 0{\cdot}125 = \sqrt{1{\cdot}125} = 1{\cdot}06$. The coefficient of relationship of B to T therefore is $(50 + 12{\cdot}5) \div 1{\cdot}06 = 58{\cdot}9$ per cent.

Collateral relationship occurs when inbreeding takes place, particularly if the matings are close. Fig. 57 is prepared to present such a case.

It is required to find the coefficient of relationship between B and A. The relation is both direct and indirect or collateral.

1. *Direct relationship*:
   - (i) Father to son
     B to B = 0 = n, B to A = 1 = n_1; = $\frac{1}{2}^1$ = 0·50
   - (ii) B related to A as sire of G
     $n = 0$, $n_1 = 3$; $\frac{1}{2}^3 = 0{\cdot}125$

2. *Indirect or collateral relationship*:
   - (iii) D sire of B also sire of F
     $n = 1$, $n_1 = 3$; $\frac{1}{2}^4 = 0{\cdot}0625$
   - (iv) N sire of E also sire of K
     $n = 2$, $n_1 = 4$; $\frac{1}{2}^6 = 0{\cdot}0156$
   - (v) N sire of E also sire of H
     $n = 2$, $n_1 = 4$; $\frac{1}{2}^6 = 0{\cdot}0156$

Examination for inbreeding:

(i) and (ii) B is inbred due to recurrence of O.
Coefficient of Inbreeding = 0·125.

(iii) D is inbred due to recurrence of L.
Coefficient of Inbreeding = 0·125.

A is inbred, Coefficient of Inbreeding = 0·281.

All the data are now available for substitution in the general formula for coefficient of relationship.

The numerator becomes:

(i) 0·50 × 1·125 (a/c B) = 0·562
(ii) 0·125 × 1·125 (a/c B) = 0·140
(iii) 0·062 × 1·125 (a/c D) = 0·069
(iv) 0·0156 = 0·015
(v) 0·0156 = 0·015

0·801

The denominator becomes:

$$\sqrt{(1 + 0{\cdot}125)(1 + 0{\cdot}281)}$$
$$= \sqrt{1{\cdot}125 \times 1{\cdot}281}$$
$$= \sqrt{1{\cdot}44}$$
$$= 1{\cdot}2$$

$$\frac{\text{Numerator}}{\text{Denominator}} = \frac{\cdot 801}{1{\cdot}2} = 66{\cdot}7 \text{ per cent}$$

Wright's coefficient of relationship between A and B, which has made allowance for the inbreeding of individuals and for collateral relationship, is therefore 66·7 per cent.

The percentage of blood relationship of A to B, shown in this pedigree is 50 per cent as a sire, and 12·5 per cent as a great-grandsire: total, 62·5 per cent.

It will be seen that, for practical purposes, and having regard to the arithmetical processes required to determine Wright's coefficient, percentage of blood relationships have much to recommend them despite their occasional minor inaccuracies. This is particularly so because (i) the two percentages are identical when the relationship is direct and uncomplicated by inbreeding, and (ii) it is desirable, when two possible matings are being considered, to determine and tabulate for adjudication all the relationships that would accrue from either mating. This magnifies the difference in the quantity of arithmetical work involved.

It is impossible to hold high degrees of relationship to particular animals without some degree of inbreeding. However, the two can be considered in relation to one another. For example, reference to examinations of pedigrees for T and V will disclose that, although T and V were both line-bred to the same animal B, the coefficient of inbreeding of T is only half that of V (12·5 per cent and 25·0 per cent). On the other hand the relationship to B, the favoured animal, was only reduced by 17 per cent in the case of T.

Inbreeding has been shown to be fraught with dangers, particularly when it is intense. The art of constructive breeding therefore is to hold high degrees of relationship, and so similarity to favoured animals, while restricting the degree and so the risks of inbreeding. Selection of procedure to do this is facilitated by calculations such as the foregoing. By them it can be determined, for example, that the system of mating that produced T is safer than that which produced V.

In practice it will be found that critical portions of pedigrees seldom extend further than four or five generations and that inbreeding has not been intense. Under these conditions, the use of percentage of blood for general relationships; of Wright's coefficient of inbreeding to determine the degree of inbreeding that would result from a mating *before it is made*; and of Wright's coefficient of relationship to determine the relationship of animals not directly related although members of the same breeding group, will be found arithmetically simple and sufficiently exact.

APPENDIX II

GLOSSARY OF TERMS

MANY of the following terms and definitions have been taken directly from the glossary of genetic terms which appeared in the United States Department of Agriculture Year-Book of 1936, pp. 153-64. Others were derived from the report of a Committee of the American Society of Animal Production. The reference to the Committee was "Terminology to Describe the Progeny of Various Systems of Breeding". Its report appeared in the Record of Proceedings of the Society's Thirty-third Annual Meeting, published in 1940, p. 378. Two definitions then reported were modified at the subsequent meeting. These have been included here. The remaining twenty per cent of the terms and definitions were prepared by the author to have special reference to the text.

Allelomorph. Mendelian characters are inherited in alternative pairs (or series). These alternative forms of a gene, which are located at the same point on each one of a pair of chromosomes, are called allelomorphs. Examples: Horns (recessive), Hornlessness (dominant). *Syn.* Allel.

Anterior pituitary. The front portion of a ductless gland at the base of the brain. It secretes chemical substances, hormones, which activate and control reproductive processes.

Atavism. Reappearance of a character after a lapse of one or more generations. (See Throwback.)

Back-cross. Cross of a hybrid or first-cross with one of the parental types.

Biometry. Application of statistical methods to the study of biological problems.

Blending inheritance. Inheritance in which dominance of either parental character is lacking. Genetic and biometrical methods have shown that this is due, in many cases, to the interaction of several multiple factors independently inherited.

Character (a contraction of characteristic). A term used to designate any form, function, or feature of an organism. The Mendelian characters of genetics represent the end products of development in which a decisive effect is produced by one or more definite genes. The constant and unchanging thing is the factor or gene itself rather than the character.

Chromatin. That part of the nuclear substance which forms the chromosomes and stains deeply with basic dyes.

Chromosomes. Dark-staining bodies visible under the microscope in the nucleus of the cell at the time of cell division. The number of chromosomes in any species is usually constant. Chromosomes carry the genes, linearly arranged, which control the development of Mendelian characters.

Class interval. A constant difference between observations when they are arranged as Frequency Tables or Graphs.

Collateral inheritance. A term used to describe the appearance of characters in collateral members of a family, as when an uncle and a niece show the same character, inherited by the related individuals from a common ancestor. Collateral inheritance is characteristic of recessive characters which appear irregularly, in contrast to dominant characters which appear regularly in successive generations.

Consanguinity. Literally, with similar blood. The adjectival form, consanguineous, is applied to matings between related animals.

Correlation. A relation between two quantities such that an increase or decrease of one is, in general, associated with an increase or decrease of the other. The degree of linear correlation is measured by the correlation coefficient, *r*, which may range in value from -1 to $+1$. When precise mathematical correlation is not known two conditions are referred to as associated, not as being correlated.

Crossbred. Progeny resulting from the mating of different breeds or varieties.

Cytoplasm. The protosplasm of a cell exclusive of the nucleus.

Cytology. The study of the structure, function and life history of the cell.

Differentiation. The process whereby cells and tissues become structurally unlike during the process of growth and development.

Dihybrid. An individual which is heterozygous with respect to two Mendelian characters.

Diploid. Having two sets of chromosomes. Body tissues of higher plants and animals are ordinarily diploid in chromosome constitution. (See Somatic.)

Distribution. A statistical term meaning the scatter of a series of observations.

Dizygotic. Originating from two fertilized eggs. Dizygotic twins are genetically no more similar than ordinary brothers and sisters.

Dominant. A character possessed by one of the parents of a hybrid which is manifested in the hybrid to the apparent exclusion of the contrasted character of the other parent (the recessive).

Endocrine. An adjective qualifying systems, secretions or glands. Hormones are the active principals of these. The Anterior Pituitary gland is that portion of an endocrine gland. (See Anterior Pituitary.)

Epigenesis. The conception of generation now accepted by biologists, which holds that the germ or embryo develops afresh in each generation through the interaction of genes and other protoplasmic components, in contrast with the theory once held that development was merely an unfolding of organs already microscopically present in the egg.

F_1 (pronounced eff-one). The first filial generation; the first generation offspring of a given mating.

F_2. The second filial generation; the first hybrid generation in which Mendelian segregation occurs (see Segregation). Produced by intercrossing the F_1. The inbred grandchildren of a given mating. F_3, F_4, etc. denote later generations from a given cross.

Factor. Same as gene.

Fertilization. The union of a male gamete with an egg, a process requisite, in the higher animals, to the development of the egg and the individual.

Frequency. A statistical term implying the number of times particular observations occur. This may be shown in Frequency Tables or Frequency Polygons. The latter are graphs derived from the data of Frequency Tables.

Gamete. A reproductive cell of either sex; a sperm or an ovum.

Gene. (i) The unit of inheritance, which is transmitted in the germ cells and which, by interaction with the genic and cytoplasmic complex and the environment, controls the development of a character. The genes are supposed to be arranged linearly in the chromosomes. (ii) "The physical basis of heredity."

Genetics. The science of heredity, variation, sex determination and related phenomena.

Genotype. The entire genetic constitution, expressed and latent in an organism, in contrast to the phenotype. (See Phenotype and Latent.)

Germ-cell. A cell capable of reproduction, or of sharing in reproduction, as contrasted with somatic or body cells which are specialized for other functions.

Germ-plasm. The material basis of heredity taken collectively.

Gonad. A reproductive gland, an ovary or testis.

Haploid. Single; referring to the reduced number of chromosomes in mature germ-cells. (See Diploid.)

Heredity. The resemblance, derived from ancestry, among organisms related by descent. All characters derived in this manner.

Heterogametic. Producing unlike gametes with respect to the sex chromosome. In the "X-Y" group of organisms, the male is the sex with different gametes (heterogametic), producing X-bearing sperms and Y-bearing sperms in equal numbers. The female (homogametic) produces only X-bearing ova. All domestic animals are of this kind. Poultry are in a Z-W group. The females are heterogametic, producing W-bearing ova and Z-bearing ova; the sperm cells all carry a Z chromosome.

Heterogeneity. From heteros, different, and genus, a kind. A state which arises when a population is made up of different kinds of individuals.

Heterosis. See Hybrid Vigour.

Heterozygote. An organism to which its two parents have contributed unlike genes with respect to any given allelomorphic pair governing contrasted characters, and which in turn produces two kinds of germ-cells with respect to the characters.

Heterozygous. Containing both genes of an allelomorphic pair or two genes of an allelomorphic series. Heterozygous individuals usually either resemble the individuals homozygous for the dominant character or are intermediate with respect to this character. They transmit the recessive gene to half their offspring.

Heterozygosity (-osis). The state of being heterozygous.

Histogram. A form of graph showing frequencies in distributions. (See Frequency and Distribution.)

Homologous. Chromosomes occur in somatic cells in pairs; the members of these pairs are spoken of as homologous chromosomes.

Homozygote. An organism whose parents contributed to it a similar member of any given pair of genes, and whose germ-cells are therefore all alike with respect to genes for that character.

Homozygous. (i) An organism formed by like germ-cells, (ii) An organism is said to be homozygous for a given character when all its germ-cells transmit identical genes for this character. The second is the most frequently used meaning of the term.

Homozygosity (-osis). The state of being homozygous.

Hybrid. (i) The offspring of two parents unlike one another in one or more heritable characters. (ii) A heterozygote. (iii) Used for the broader categories for example, numerical hybrid, one whose parental gametes differed in number of chromosomes; structural hybrid, one whose parental gametes differed in the structure of their chromosomes.

Hybrid vigour. The superiority over the better parent that is exhibited by the progeny. This applies to progeny resulting from the crossing of strains, varieties, breeds or species.

Hybridization. The process of crossing organisms of unlike hereditary constitution, thereby producing F_1 offspring possessing genes for the traits of both parents.

Identical twins. Two individuals developed from the same fertilized egg. They are extremely alike and are generally (but not always) found in the same membranes before birth. Also referred to as monozygotic twins.

Inbred line. The progeny resulting from breeding closely-related animals. To be classed as an inbred strain, the individuals should have an average coefficient of inbreeding of 37·5 per cent, equivalent to two generations of brother-sister matings.

Incross. Progeny resulting from the crossing of individuals of inbred lines within the same breed or variety.

Incross-bred. Progeny resulting from the crossing of individuals from inbred lines of different breeds or varieties.

Inseminate. To introduce semen into the female genitalia. It is effected naturally by coitus into either the uterus (mare) or vagina (sheep and cattle).

Latent. A term applied to traits or characters whose genes exist in the germ-plasm of an organism although the traits are not apparent in the organism. (See Dominant.)

Lethal factor. A factor which renders an organism expressing it unable to live. Inheritable lethal mutations are recessive in their lethal effect and must be homozygous to be fatal to the organism.

Linkage. Association of genes in inheritance due to the fact that they are located in the same chromosome.

Maturation. A process which germ-cells undergo before they become functional. It consists usually of two cell divisions, during which the number of chromosomes is reduced to one-half that of the body-cells.

Mendel's law. Mendel enunciated three basic principles of heredity: (i) Inheritance of character determiners (genes) as allelomorphic units, not as a blend: (ii) dominance; (iii) independent random segregation of allelomorphs in the formation of gametes and their random recombination at fertilization. As a consequence of these conditions a given allelomorphic pair of unit characters appears in a certain definite predictable ratio (3:1; 1:2:1; 1:1) among the progeny of a given mating.

Micron. Measurement of microscopic objects requires a minute unit. This is the micron, equal to the millionth of a metre.

Monofactorial. Controlled by a single gene difference (factor) in inheritance.

Monohybrid. The offspring of two parents differing in only one Mendelian characteristic.

Monozygotic. Originating from a single fertilized egg which in the course of subsequent development has given rise to two or more individuals. Twins (or triplets, etc.) of such origin are genetically identical. (See Identical Twins.)

Mosaic. An organism part of which is made up of tissue genetically different from the remaining part. Mosaics are produced by mutations during development or by aberrant chromosome behaviour.

Multiple allelomorphs. The existence of more than two alternative forms of a gene. Any two members of a series of multiple allelomorphs may be allelomorphic to each other. A single individual having only the normal number of chromosomes can bear only two members of such a series.

Multiple factors. Two or more pairs of factors having a similar supplementary, complementary, or cumulative effect. In a broader sense, any factors working together to produce a single result. (See Polygenic.)

Mutation. A sudden variation that is later passed on through inheritance and that results from changes in a gene or genes.

Nicking. When individuals or strains of animals produce better progeny when mated with some strains than with others, those that give rise to the superior progeny are stated to have nicked. The phenomenon probably depends upon maximal complementary action of genes from one strain with those of the other.

Nucleus. A highly refractive deeply-staining body of specialized protoplasm found within nearly all cells. The chromosomes are found in the nucleus.

Oögenesis. The series of changes, including one or two divisions, which the reproductive cells of female animals undergo before they are ready for reproduction.

Outcross. A cross of relatively unrelated animals within the same breed or variety.

Ovum. An egg.

Partial dominance. Capability of being produced to an intermediate degree when only one member of a pair of genes is present.

Phenotype. The organism as exemplified by its expressed characters as contrasted with its genetic constitution. (See Genotype.)

Polar bodies. Two minute cells cast off from the animal ovum during maturation.

Polygenic. Having many genes. Adjectival synonym for multiple-factor when applied to a character or method of inheritance. (See Multiple Factors.)

Population. Used genetically and statistically to indicate collectively all individuals or units.

Prepotent. A sire is prepotent when many or all of his progeny are similar to him. Prepotency depends principally upon dominance of the character, homozygosis of the sire and a suitable contribution by the female. (See also Dominance, Homozygosis, and Character.)

Protoplasm. "The physical basis of life"; a chemical compound or probably an emulsion of numerous compounds. It contains proteins which differ slightly in each species of organism. It contains carbon, hydrogen, oxygen, nitrogen, sulphur, and various salts, but is so complex that it defies structural analysis.

Pure line. A strain of organisms that is genetically pure (homozygous) because of continued inbreeding or through other means.

Race. A group of individuals having certain characteristics in common because they have common ancestry–generally a subdivision of species.

Random assortment. Occurring by chance.

Recessive. See Dominant and Latent.

Reciprocal. Involving the same types of individuals but with the sexes reversed; said of two crosses in one of which the female possesses the same characteristic as does the male in the other cross.

Reduction division. The division preceding the formation of gametes in which the chromosome pairs separate and the chromosome number is reduced to half the somatic number.

Replicates. Repetitions, generally of experiments to establish accuracy of observations.

Reversion. Appearance of a character possessed by a remote ancestor but not by recent ones. (See Atavism and Throwback).

Secondary sexual character. A character dependent for its expression upon chemical substances from the male or female gonads but having no direct reproductive function.

Segregation. Separation of the genes of a pair in the formation of germ cells, the two genes going to different gametes. Segregation is manifested in F_2 (and later hybrid generations) as a separation and distribution to different individuals of the Mendelian characters in which the parents of the F_1 hybrid differed. Often referred to as Mendelian segregation.

Semen. The fluid emitted by males at coitus. It is derived from accessory glands and contains spermatozoa. (See Sperm.)

Sex linkage. Association of an hereditary character with sex due to the fact that its gene is on a sex chromosome.

Sex limited. Appearance (or expression) of an inherited character in only one sex due either to sex linkage or to differential dominance in the two sexes.

Sibling. One of two or more children of the same parents but not of the same birth and without regard to sex.

Skewed. Normal distributions are bell-shaped. Graphs of other types are distorted and slant to one side. The curve is then said to be skewed. (See Distributions and Frequency.)

Somatic. Referring to body tissues; having two sets of chromosomes, one set normally coming from the female parent and one from the male; as contrasted with mature germ-cells having a single set of chromosomes.

Sperm. The spermatozoon (plural: -zoa), sperm-cell or germinal cell of the male.

Spermatogenesis. The formation of male germ-cells.

Sport. An abrupt deviation from type. (See Mutation.)

Telegony. A disproven conception that the nature of the male parentage of a dam's previous offspring can affect those conceived and borne at a later date to a different sire.

Top-cross. The mating of a male of a certain family to females of another family of the same breed or variety.

Top-incross. Progeny resulting from the cross of inbred sires on non-inbred dams of the same breed or variety.

Top-crossbred. Progeny resulting from the mating of inbred sires with non-inbred dams of different breeds or varieties.

Throwback. Progeny unlike the parents but similar to a remote ancestor. (See Atavism.)

Trait. A term often used as a synonym of "character".

Trihybrid. A hybrid heterozygous with respect to three pairs of Mendelian factors.

Two-factor inheritance. Some characters depend for their development on two hereditary elements (genes) both of which must be present to produce an effect.

Unit character. An hereditary trait that behaves as a unit in transmission, being capable of inheritance independently of other unit characters and determined by a single gene pair.

Variation. In biology, the occurrence of differences among individuals of the same species or variety.

Variety. A division of a species; a group of individuals within a species that differ in some minor respect from the rest of the species.

Virus. A disease-producing substance in plants and animals. Its particles are ultra-visible even with the strongest microscopes and pass through the finest of filters.

Zygote. The cell produced by the union of two cells (gametes) in reproduction; also the individual developing from such a combined cell.

APPENDIX III

REFERENCES TO LITERATURE

CITATIONS made with the year of publication and the title of the book or article have particular reference to the part of the text in which they have been quoted. However, the admirable breed histories prepared by Sinclair and Macdonald, by Youatt, by Plumb, and the historical reviews by Slater and by Roberts, have been used so widely that individual references in so many instances were impossible. The author cannot pay greater tribute to these texts than to state his indebtedness to them and highly recommend them as works of reference to other readers.

Those who wish to explore a wider field of literature than is stated will find that all the books and most of the articles quoted provide other references. Collectively they will disclose a range of reading far more comprehensive than could possibly be included here.

Ahmed, I. A., "The Structure and Behaviour of the Chromosomes of the Sheep During Mitosis and Meiosis", *Proceedings of the Royal Society of Edinburgh,* vol. lx, Part III, no. 20, 1939.

Babcock, E. B., and Clausen, R. E., *Genetics in Relation to Agriculture.* McGraw-Hill Company, New York, 1927.

Barker, S. G., *Wool, A Study of the Fibre.* H.M. Stationery Office, London, 1929.

Calder, A., "The Role of Inbreeding in the Development of the Clydesdale Breed of Horses", *Proceedings of the Royal Society of Edinburgh,* vol. xlvii, Part II, no. 8, 1926.

Daley, C. J., and Eastoe, R., "Breeding Habits of the Dorset Horn", *Agricultural Gazette of New South Wales,* vol. liv, no. 2, 1943, p. 75.

Dobzhansky, Theo., *Genetics and the Origin of Species.* Columbia University Press, 1941.

East, E. M., and Jones, D. F., *Inbreeding and Outbreeding.* The Lippincott Company, Philadelphia, 1919.

Eaton, N. D., "Effect of Crossing Inbred Lines of Guinea Pigs", Bureau of Animal Industry, *Technical Bulletin,* no. 765, 1941.

Edwards, J., "The Progeny-Test as a Method of Evaluating the Dairy Sire", *Journal of Agricultural Science,* vol. xxii, 1932, p. 811.

Fourie, P. J. J., "Bovine Congenital Porphyrinuria (Pink Tooth) Inherited as a Recessive Character", *Onderstepoort Journal of Veterinary Science,* vol. xiii, pp. 383-98, 1939.

Goodale, H. D., "Breeding to Meet Economic Needs: Methods Applicable to the Improvement of Livestock", *American Naturalist,* vol. lxxii, 1938, p. 243.

Goodrich, E. S., *The Evolution of Living Organisms.* T. C. and E. C. Jack, London, 1912.

Grant, Macpherson C., *History of the Ballindalloch Herd of Aberdeen-Angus Cattle 1861-1891.* Banffshire, 1892.

Graves, R. R., and Fohrman, M. H., "Superior Germ Plasm in Dairy Herds", United States Department of Agriculture, *Yearbook of Agriculture,* 1936, pp. 997-1069.

Gunn, R. M. C., "Fertility in Sheep. Artificial Production of Seminal Ejaculation and the Characters of the Spermatozoa Contained Therein", Commonwealth of Australia, Council for Scientific and Industrial Research, *Bulletin,* no. 94, 1936.

Gunn, R. M. C., Sanders, R. N., and Granger, W., "Studies in Fertility in Sheep. Seminal Changes Affecting Fertility in Rams", Commonwealth of Australia, Council for Scientific and Industrial Research, *Bulletin,* no. 148, 1943.

Hagedoorn, A. L., *Animal Breeding.* Crosby Lockwood and Son, London, 1939.

Hammond, J., *Farm Animals. Their Breeding, Growth, and Inheritance.* Edward Arnold and Company, London, 1940.

Hammond, J., "British Animal Husbandry", *Endeavour,* vol. i, Part IV, I.C.I., London, 1942.

Hazel, L. N., and Lush, Jay L., "The Efficiency of Three Methods of Selection", *Journal of Heredity,* vol. xxxiii, 1942, pp. 393-9.

Jenkins, M. T., "Corn Improvement", United States Department of Agriculture, *Yearbook of Agriculture,* 1936, pp. 455-552.

Kelley, M. A. R., and Rupel, I. W., "Relation of Stable Environment to Milk Production", *United States Department of Agriculture Technical Bulletin,* no. 591, 1937.

Kelley, R. B., "Studies in Fertility of Sheep", Commonwealth of Australia, Council for Scientific and Industrial Research, *Bulletin,* no. 112, 1937.

Kelley, R. B., *Sheep Dogs. Their Breeding, Maintenance and Training.* Angus and Robertson Limited, Sydney, 1946.

Kelley, R. B., "Fertility in Sheep. An Experimental Study of Periodicity of Oestrus and Non-breeding Seasons in Australia", Commonwealth of Australia, Council for Scientific and Industrial Research, *Bulletin,* no. 166, 1943.

Kelley, R. B., "Zebu-cross Cattle in Northern Australia: An Ecological Experiment", Commonwealth of Australia, Council for Scientific and Industrial Research, *Bulletin,* no. 172, 1943.

Kelley, R. B., and Shaw, H. E. B., "Note on the Occurrence and Inheritance of Pigmented Wool", Commonwealth of Australia, *Journal of the Council for Scientific and Industrial Research,* vol. xv, no. 1, 1942; vol. xvi, no. 1, 1943.

Lambert, W. V., and McKenzie, F. F., "Artificial Inseminations in Livestock Breeding", United States Department of Agriculture, *Circular,* no. 567, 1940.

Lewis, A. A., "Seventy Years of Selection for Conformation in Dairy Cattle", *Journal of Dairy Science,* vol. xxv, 1942, pp. 489-501.

Lindstrom, E. W., "Analysis of Modern Maize Breeding Principles and Methods", *Proceedings of the 7th International Conference on Genetics.* Cambridge University Press, 1939.

Locy, Wm. A., *Biology and Its Makers.* Henry Holt and Company, New York, 1915.

Lush, Jay L., *Animal Breeding Plans.* Collegiate Press, Ames, Iowa, 1937.

Lush, Jay L., "The Use Which Has Been Made of Inbreeding in Establishing and Improving Pure Breeds", *Proceedings of the International Congress of Zootechnology,* Zürich, 1939.

Lush, Jay L., "Intra-Sire Correlations or Regressions of Offspring on Dam as a Method of Estimating Heritability of Characteristics", *33rd Annual Proceedings of the American Society of Animal Productivity,* 1940.

Manresa, M., and others, "The Influence of Atmospheric Temperature upon Haemoglobin and Other Constituents of the Blood of Cattle", *Proceedings of the 7th International Conference on Genetics.* Cambridge University Press, 1939.

Mather, K., "Genetics and the Russian Controversy", *Nature,* vol. cxlix, p. 3781.

Macdonald, J., and Sinclair, J., *History of Aberdeen-Angus Cattle,* edition revised by Jas. Sinclair. Vinton and Company Limited, London, 1910.

Macdonald, J., and Sinclair, J., *History of Hereford Cattle,* edition revised by Jas. Sinclair. Vinton and Company Limited, London, 1909.

McMeekan, C. P., "Growth and Development in the Bacon Pig with Special Reference to Carcase Quality", *Journal of Agriculture Science,* vol. xxx, Parts II-IV, 1940; vol. xxxi, Part I, 1941.

McPhee, H. C., "Recent Attempts to Co-ordinate Genetic Research on Farm Animals in the United States", *Proceedings of the 7th International Conference on Genetics.* Cambridge University Press, 1939.

McPhee, H. C., and Hankins, O. G., "Swine: Some Current Breeding Problems", United States Department of Agriculture, *Yearbook of Agriculture,* 1936, p. 887.

McPhee, H. C., and Spencer, D. H., "Breeding Problems with Sheep", United States Department of Agriculture, *Yearbook of Agriculture,* 1936, p. 907.

Nichols, J. E., *Livestock Improvement in Relation to Heredity and Environment.* Oliver and Boyd, Edinburgh, 1944.

Phillips, R. W., Black, W. H., Knapp, B., and Clark, R. T., "Crossbreeding for Beef Production", *Journal of Animal Science,* vol. i, no. 3, 1942, pp. 213-19.

Plumb, Chas. S., *Types and Breeds of Farm Animals.* Ginn and Company, New York, 1920.

Prentice, E. P., *Breeding Profitable Dairy Cattle.* Houghton Mifflin Company, Boston, 1935.

Prentice, E. P., *American Dairy Cattle: Their Past and Future.* Harper and Brothers, New York, 1942.

Regan, W. M., and Richardson, G. A., "Reactions of the Dairy Cow to Changes in Environmental Temperature", *Journal of Dairy Science,* vol. xxi, Part II, 1938, pp. 73-9.

Rhoad, A. O., "A Scale of Heat Tolerance for Cattle", *Journal of Animal Science,* vol. i, no. 1, 1942, p. 85.

Ritzman, E. G., "Inheritance of Size and Conformation in Sheep", *Technical Bulletin,* no. 25, 1923, of the New Hampshire Agricultural Experimental Station, U.S.A.

Ritzman, E. G., and Davenport, C. B., "Comparison of Some Traits of Conformation of Southdown and Rambouillet Sheep", *Technical Bulletin,* no. 15, 1920, of the New Hampshire Agricultural Experimental Station, U.S.A.

Roberts, D. W., *An Outline of Economic History of England to 1939,* fourth edition. Longmans, Green and Company, London, 1940.

Roberts, S. H., *The Squatting Age in Australia, 1835-1847.* Melbourne University Press, 1935.

Shaw, A. M., and MacEwan, J. W. G., "An Experiment on Beef Production in Western Canada", *Scientific Agriculture,* vol. xix, 1938, pp. 177-98.

Shier, F. L., and Underwood, E. J., "Studies in Sheep Husbandry in Western Australia. I. The Effect of Nutritional 'Flushing' upon Fertility" *Journal of the Department of Agriculture,* Western Australia, vol. xviii, no. I, 1941.

Sinclair, J., *History of Shorthorn Cattle.* Vinton and Company, London, 1908.

Slater, C., *The Growth of Modern England,* second edition. Constable and Company, London, 1939.

Smith, A. D. Buchanan, "A Statistical Enquiry into the Inheritance of Milk Yield in Three Herds of Dairy Shorthorn Cattle", *Journal of Dairy Research,* vol. viii, no. 3, 1937.

Stonaker, H. H., Holbert, J. C., and Lush, Jay L., "The Breeding Structure of the Pure-bred Aberdeen-Angus Cattle Population in the United States from 1900 to 1939", *Journal of Animal Science,* vol. i, no. 4, 1942, p. 343.

Thompson, D. S., "A Study of the Breeding Seasons of a Group of Merinos and Three Groups of Half-breds at Roseworthy Agricultural College", *Journal of Agriculture,* South Australia, April, 1942.

Verges, J. B., "Effect of the Plane of Nutriment on the Carcase Quality of Suffolk Cross Lambs", *Yearbook* of the Suffolk Sheep Society, England, 1939.

Watson, J. A. Scott, "The Danish Agricultural Exhibition, 1938", *Transactions* of the Highland Agricultural Society of Scotland, 1939, pp. 1-15.

Wentworth, Lady, *Thoroughbred Racing Stock.* George Allen and Unwin Limited, London, 1938.

Wilson, Jas., *The Evolution of British Cattle.* London, 1909.

Woodworth, Robt. S., *Psychology.* Methuen and Company, London, 1934.

Wright, S., "Systems of Mating. The Biometric Relations Between Parent and Offspring", *Genetics,* vol. vi, no. 2, 1921.

Wright, S., "Mendelian Analysis of Pure Breeds of Livestock. II. The Duchess Shorthorns", *Journal of Heredity,* vol. x, no. 9, 1923.

Wright, S., and McPhee, H. C., "An Approximate Method of Calculating Coefficients of Inbreeding and Relationship from Livestock Pedigrees", *Journal of Agricultural Research,* vol. xxxi, no. 4, 1935.

Youatt, Wm., *Sheep: Their Breeds, Management and Diseases.* Simpkin, Marshall and Company, London, 1837.

Youatt, Wm., *Cattle: Their Breeds, Management and Diseases.* Simpkin, Marshall and Company, London, 1874.